The Science of Grace!

Ifeanyi Chukwujama

This Book is available at:
www.amazon.com
www.kindle.com
www.createspace.com
www.Jesus-On.com
www.inchristwetrust.org
www.truthnlife.com

Copyright © 2015 Ifeanyi Chukwujama

All rights reserved.

For permission to reprint or copy this booklet, please contact the publisher:

Ifeanyi Chukwujama

Truth & Life Institute
Incorporated

394 Warren Street, Boston, MA 02119

Email: JesusOnkm@gmail.com

Website: www.Jesus-On.com

ISBN-10: 1511826150

ISBN-13: 978-1511826150

DEDICATION

To my children and my wife! I thank God Almighty for allowing me to live long enough to teach my children that the most important possession in this life is having an intimate relationship with God and His Son Jesus Christ through obedience and complete surrender of our lives to Jesus Christ!

Also to my readers who come to share the same value in Jesus Christ! May our merciful and loving God maintain His perpetual love in our hearts, and keep us under His glorious wings, forever and ever. Amen!

CONTENTS

	Acknowledgments	i
1	The Science of Grace	Pg. 1
2	The Threshold of Life	Pg. 51
3	There Are More Particles than We Know	Pg. 67
4	Only You Can Let The Devil into Your Life	Pg. 81
5	Grace Is Beyond Religion!	Pg. 89
6	Religion May Hamper True Worship of God!	Pg. 101
7	The Power of Grace	Pg. 145
8	The World has not seen a Single Religious War	Pg. 153
	About the Author	Pg. 177
	Other Titles from This Author	Pg. 179

ACKNOWLEDGMENTS

The only teacher any of us needs in this world is God. His word imprints on your heart as a tattoo; and lives inside of you throughout your life.

We are living in the end times when God promises to write His laws on the hearts of men so that no one needs to say to his brother, "Do you know the Lord?" Because we will all know God and revel in his boundless love and unending joy.

Desire the Lord with all your heart, and all your soul and all your mind; and you will activate from within you, the rivers of living water—God's Grace; and live in holiness and the manifest powers of His Spirit.

The Bible is the source of all knowledge. Embrace it and live!

CHAPTER ONE

The Science of Grace

"One with God is majority," used to be a popular saying when majority of the people in the civilized world still had faith in God. Those were the days when innocence still existed in the world; and the people, and their leaders, believed in a common destiny in Jesus Christ.

Then came the explosive speed of science—the explosiveness which changed the way the people in the civilized world think and operate. Innocence, then, gave way to the irreversible assertion of the human mind over the human spirit.

Empiricism, reasoning and logic, forced spirituality into antiquity, where only the broken and the heavily-trodden-over dare to retrieve it, and dust it off; for salvage value they hope could piece their broken lives back together again. Thus, they renew their hearts and their spirits!

Yet, those who championed the elevation of the human mind over the human spirit, and made invalids out of common men and women, continue to take credit for everything. Credit, even for the rehabilitation of those who were victims of the austere economic measures that marginalized and sidelined them in the first place!

Civilized societies now resort to social programs to cover over the exclusive financial competitiveness they have created and fostered in which only the fittest—the most brutish—can survive.

The age of philanthropy dawned on the world. Governments, corporations and wealthy individuals have got a new cover for greed. The giving has some magic to it—even miraculous! It glosses over societal ills. As long as there is giving, the method of accumulating the wealth in the first place, no longer matters. Meteoric rise to fame and fortune overnight has become the order of the day.

Every corporation sets up a philanthropic foundation to achieve this purpose of earning wealth fast and given back some. Even athletes are required by their contracts to engage in philanthropy so that the large amounts of money reported through these philanthropic activities would be used to justify greedy and unfair economic policies and practices.

Philanthropy has become the new spirituality. As long as you give back some, you are an outstanding citizen. Observing the commands of God no longer seems relevant, especially since those commands seem to be slowing down profit making.

The new attitude is to capture wealth as fast as possible, and quickly move on to greater and yet more profitable economic and financial schemes. The stock market and its related specialties add more legitimacy to fast and immoral grabs, and concentration of wealth in fewer hands. All the disenfranchised are only sustained by the Grace of God, at the same time social programs are credited for carrying them.

Organized philanthropy is valuable, aside from providing a cover for greed and wickedness. But philanthropy—the way the world governments, rich corporations and the wealthy social elites, use it—will never become a substitute for spirituality.

Everything we do in life is done through Grace. Grace enables; and Grace monitors and maintains the track records of every human person, government or organization. Grace is in all things; Grace is around all things; and Grace connects all things to one another. Grace never misses a beat.

Therefore, everything any human person, any human government or any human organization, does is captured in real time by Grace. And each and every step of what transpired—from the very beginning to the very end—is maintained by Grace; and rewarded or punished accordingly.

And Grace never seeks inputs from any person, any government or any organization, before Grace takes action, in favor of, or against anything; or anybody.

Grace does everything Grace does, out of Grace's own determination and Grace's timing. Grace only serves God's purpose, according to God's design and God's predestination. And when it is all said and done, the only thing that remains is Grace, because Grace is God the Father, God the Son and God the Holy Spirit!

The world's philanthropy may be generating great buzz, but it does not impress God. Philanthropy within Grace is the only God-sanctioned philanthropy. It is not the size of philanthropy that matters to God. It is how the wealth used in philanthropy is generated, and how it is dispersed, that matter to God.

What God sanctioned is that societies create for all citizens, regular and dignified access to good source of income—and not to subject a large segment of society to dependency on charity and philanthropic giving, as a means of meeting their daily life's necessities.

To create welfare conditions in the first place, and then turn around and tackle the problems through social programs, is the arrogance of the devil. These programs as constantly touted as signs of society's absolute humanity, its benevolence and its fairness. Is it really? Or could society bring more parity, with dignity, to the lives of all its citizens?

Even the church is falling far behind in helping those in the kingdom of God maintain dignified access to income. The church has virtually relegated the job of income-creation to governments and corporations, absolving itself from its foremost responsibility—to feed the people of God. Here is Jesus Christ Himself, commanding Apostle Peter to fulfill the daily needs of all God's people:

"When they had finished eating, Jesus said to Simon Peter, "Simon son of John, do you love me more than these?"

"Yes, Lord," he said, "you know that I love you."

Jesus said, "<u>Feed my lambs</u>."

[16] Again Jesus said, "Simon son of John, do you love me?"

He answered, "Yes, Lord, you know that I love you."

Jesus said, "<u>Take care of my sheep</u>."

[17] The third time he said to him, "Simon son of John, do you love me?"

Peter was hurt because Jesus asked him the third time, "Do you love me?" He said, "Lord, you know all things; you know that I love you."

Jesus said, "<u>Feed my sheep.</u>" (John 21:15-17)

And did Peter and his brother apostles comply with that command? Not only did they fulfill it, they excelled in it! The apostles took this command so seriously that when it came time for them to demonstrate that they are equal to the task, they performed flawlessly. Here is that passage from the Bible:

"They devoted themselves to the apostles' teaching and to fellowship, to the breaking of bread and to prayer. ⁴³ Everyone was filled with awe at the many wonders and signs performed by the apostles. ⁴⁴ <u>All the believers were together and had everything in common. ⁴⁵ They sold property and possessions to give to anyone who had need.</u> ⁴⁶ Every day they continued to meet together in the temple courts. They broke bread in their homes and ate together with glad and sincere hearts, ⁴⁷ praising God and enjoying the favor of all the people. And the Lord added to their number daily those who were being saved." (Acts 2:42-47)

With growth came newer and greater challenges in maintaining the welfare of the entire group of believers in Jesus Christ. Yet, Apostle Peter and his brother apostles did not feel overwhelmed and give up the charge from Jesus Christ to feed all His believers.

Calmly and confidently, and with unparalleled resolve, they decided to pass the new challenges over to those among their number, who have enough faith and Grace in them to resolve the problems.

Led by Grace, the congregants selected seven wise and discerning men to head the project of ensuring that every believer's needs were met. And the selected group performed their new appointments satisfactorily, and the church greatly increased in size and in God's favors. Here is that passage from the Bible:

"In those days when the number of disciples was increasing, the Hellenistic Jews among them complained against the Hebraic Jews because their widows were being overlooked in the daily distribution of food. ² So the Twelve gathered all the disciples together and said, "It would not be right for us to neglect the ministry of the word of God in order to wait on tables. ³ Brothers and sisters, choose seven men from among you who are known to be full of the Spirit and wisdom. We will turn this responsibility over to them ⁴ and will give our attention to prayer and the ministry of the word."

⁵ This proposal pleased the whole group. They chose Stephen, a man full of faith and of the Holy Spirit; also Philip, Procorus, Nicanor, Timon, Parmenas, and Nicolas from Antioch, a convert to Judaism. ⁶ They presented these men to the apostles, who prayed and laid their hands on them.

⁷ So the word of God spread. The number of disciples in Jerusalem increased rapidly, and a large number of priests became obedient to the faith." (Acts 6:1-7)

There were cultural, economic, political, ideological and geographic distinctions among believers from many different nations who made up the church, but not even those differences were enough to dampen their resolve; nor did the differences prevent them from coming up with effective measures to continue to live and share in love and in peace, inside Grace.

That is the great hallmark of Grace. In the kingdom of God, increased life challenges translates into increased Grace and wisdom—unlocked by the faith of the believers. That was the very reason why the unschooled Apostle Peter and his apostle brothers were able to handle these complex economic issues like experts and well-experienced professionals. Grace brought them together and Grace kept them together as one people.

It was not some great education of their minds that permitted that first century church to perform at this highest level of competence. It was the enrichment of their spirits by Grace that gave them the wisdom to operate effectively and efficiently— within a short timeframe— to the capacity which Jesus Christ commanded them to. And they accomplish this without the apostles taking any time away from their main commission—spreading the good news of the kingdom of God to all the world.

But somehow, the current-day church has abandoned Grace and Grace's glorious powers; and instead settled on the power of the mind (education)—in much the same way as the rest of the world—causing the world to, now, look at the glorious achievements of the apostles and the early church as fairy tales.

Therefore, the world's disregard, and disrespect, of God and the Bible—which prevented mankind from learning about Grace as real and substantive—is a failing of the church. The church backed away from the power of Grace, but instead, competes with the world on who can use the power of the mind better—even in matters of the spirit.

We are all disappointing God and His Christ, and depriving ourselves of the great power of Grace that is continually available to those who have faith in God. God and all the people in the family of God must not be treated as brands.

When every church denomination or group feels contented with simply catering to its direct membership alone, that denomination or group is treating the family of God as a conglomerate of brand. That is sinful!

God is one and the family of God, the world over, is one. The fellowship of the Holy Spirit all over the world is one, because that fellowship is accomplished through the one Grace that is abundantly available throughout the world and all over the universe.

Every part of God represents all of God. And every member of God's family represents all of God's family and God. Each member of God's family has the entire power of God behind him or her.

If you truly believe in Jesus Christ and obey God fully, all the powers of God is at your disposal anytime and every time. The Bible assures us that the same power with which God raised Jesus Christ from the dead—that awesome power of God—is available to each and every one of us that have sufficient faith in God and His Son Jesus Christ. This power does not work according to the will of any human being. This power works according to the will of God, and for the glory of God.

Why are Christians treating the word of God as brands? And why are Christians non-repentant in this sin? Having different family names within the one Grace is acceptable, because we all are given different names. But operating exclusively within that family name—without sufficient care and considerations for all other members of God's family—is a transgression.

All members of God's family belong together, irrespective of their different family names; and must recognize and embrace all the other members of God's family on the earth. That is what the Great

Commission was intended to accomplish—unity of God's people the world over.

The word of God is the indisputable truth of God. The word of God is about the one Christ—the truth, the light, the life and the way to God! And the word of God has only one flavor. That one flavor is the clean and flawless truth of God. And this truth was established before the foundations of the world.

Therefore, why are Christians still bent on branding the word of God and wearing their different brands of the gospel as badges of honor? Is there any honor in the disobedience of God? There is only one plausible answer to the practice of branding the gospel of Jesus Christ.

That answer: Christians do not really understand the Grace of God. Grace is the life-giving Spirit who is Jesus Christ, the Son of God. Grace extends from the Spirit of God; making Grace, the Spirit of God!

For anything to have life, that thing must have sufficient capacity to absorb and retain preset levels of Grace within it. This preset level of Grace is the threshold of life for each class of living things.

When Grace is reduced below the threshold of live in any living being; that living being dies. Anything that is devoid of Grace simply does not exist. Grace is the substance of all existence!

Figure 1: The Science of Grace - Grace transverses the entire earth and the entire universe in unbroken waves that travels at the "Speed of God". Grace travels at the speed of God because Grace is God the Father, God the Son and God the Holy Spirit. Grace is the good particle of God—the one particle that feeds and replenishes; monitors and controls; and counteracts and overrides; all other particles.

Similarly, everywhere there are evil beings, there is inactive evil; proportional to their evil. Mankind, through sin, activates the evil; and evil rots and degrades the human life, and the body faculties and their associated life processes.

Grace is the very life that is in all human beings, therefore Jesus Christ who is Grace is the Life that exists inside of every human being. When He puts more Grace inside you, He is actually putting more of Himself inside you, so that you may have life more abundantly. And Grace, who is Jesus Christ, is the God of creation. That is why Jesus Christ declared in the Bible:

"I and the Father are one." (John 10:30)

Grace is practically everything and everybody that exists on the earth and in the universe. Grace shapes and molds everything in existence into what it has become. And Grace is the power that runs everything there is. That is the reason the Bible declares:

"There is one body and one Spirit, just as you were called to one hope when you were called; ⁵ one Lord, one faith, one baptism; ⁶ <u>one God and Father of all, who is over all and through all and in all.</u>

⁷ <u>But to each one of us grace has been given as Christ apportioned it.</u>" (Ephesians 4:4-7)

And a few verses later, the Bible says:

"He who descended is the very one who ascended higher than all the heavens, in order to fill the whole universe." *(Ephesians 4:10)*

And like energy in natural and life processes, Grace is consumed in all natural and life processes. So, to sustain life or viability in any natural or supernatural body or system, Grace needs to flow continually into the body or system, to replenish whatever is used up in the natural or life processes which Grace is supporting.

Grace is the unfathomable force that held the earth and the water together at the start of creation, and kept both suspended over nothing as God complete His masterful creations *(Genesis 1:2)*. Today, Grace continues to hold everything in existence together.

Grace is the ionizing light that dawned on the earth on Day 1 of creation, when God summoned light to light up the earth *(Genesis 1:3-5)*. Today, Grace continues to feed the sun and all the stars and quasi in the universe.

Grace is the intricately explosive and finely tuned force that created space over the earth, and pushed out the greater part of the water of creation which was in the hollow of God's hand to the outer perimeter of space *(Genesis 6-10)*. Through Grace, God, on Day 2 of creation, established the complete blue print of the universe—the orbits, orbital configurations, orientations and speeds—in preparation for the creation of the universe on Day 4 of creation.

Grace is the explosive and obliterating energy with

which God brought land out of the waters in the hollow of His hand *(Genesis 1:9-11)*. Today, Grace continues to spit molten lava out from the churning core of the earth to add to land and to create new land out of the sea.

Grace is the gentle breath of life that planted vegetation throughout the surface of the earth later on Day 3 of creation and caused them to bloom *(Genesis1:11-13)*. Today, Grace continues to renew and reinvigorate life in all plant on the earth.

Grace is the cataclysmic explosive force that that scattered flaming oceans of fire and obliterating energies across space on Day 4 to create the universe *(Genesis 1:14-19)*. Today, Grace continues to add vitality to all the stars and the quasi throughout the universe in real time, so they can sustain their respective intensities.

Grace is the blue print of life that started animal lives in the seas and the atmosphere of the newly formed earth on Day 5 of creation, making the seas teem with fish and sea creatures of various kinds, and the air teem with birds of various kinds *(Genesis 1:20-23)*. Today, Grace continues to renew and replenish the life processes of these creatures.

Grace is the breath of life that, on Day 6 of creation, formed the living creatures that live on land in their various forms and filled the earth with them *(Genesis 1:24-25)*. Today, Grace continues to renew and replenish the life processes of these creatures.

Grace is the awesome power, and infinitely loving force, that shaped mankind from wet mud and stays inside mankind to maintain life in them *(Genesis 1:26-30)*. And Grace is the favor of God which was given to mankind right after mankind was created to guide mankind through natural life, and into eternal life. Today, Grace continues to renew and replenish the life processes of every human being, whether they obey God or not.

Grace is the freedom to choose which mankind received at birth so mankind could chose to obey God or not to obey God. But mankind decided to choose the organic hardware that is the human mind over the human spirit—who is the Grace that lives inside each and every human person. Today, Grace allows each human person to continue to exercise that choice, constantly nudging at every one of us to choose the directions of the human spirit over the directions of the human mind.

The human science has a gross misunderstanding of life. It has been treating life processes as if they were life itself. That is a gross misrepresentation of life. Life transcends biology and any other science. Human science cannot capture life and dissect life so it could study and understand life, because life is Grace who is God the Father, God the Son and God the Holy Spirit. That is why the Bible says that God after God molded us from wet mud, God put His breathe into us—His awesome Spirit—and we became living beings.

Throughout human existence on the earth, mankind has always looked at the body, the biological processes and the faculties inside of us, as who we are. That is why the Bible authoritatively declared as follow:

*"You say, "**Food for the stomach and the stomach for food**, and God will destroy them both." **The body**, however, is not meant for sexual immorality but **for the Lord, and the Lord for the body**. ¹⁴ By his power God raised the Lord from the dead, and he will raise us also. ¹⁵ **Do you not know that your bodies are members of Christ himself?**" (1 Corinthians 6:13-15)*

"Food for the stomach and the stomach for food:" This is a prevailing biological understanding—an understanding that has been held by mankind even before Jesus Christ came into the world more than two thousand years ago. These days we say: *We eat to live and we live to eat!*

This scientific knowledge is not new. We have added more details, but the concept is still the same. And the Bible took time to correct that understanding of science that the human body is simply some conflagration of biological systems designed to sustain human life. That notion of science is fundamentally flawed.

In correction, the Bible then declared:

"The body was meant for the Lord, and the Lord for the body (1 Corinthians 6:13).

Adding:

"Do you not know that your bodies are members of Christ himself? (1 Corinthians 6:15)."

1 Corinthians 6:13 is saying that the bodies of all human beings are, collectively, extensions of the body of Jesus Christ. Jesus Christ is Grace! And since our bodies as parts of Jesus Christ Himself, our bodies are therefore 100% Grace.

The anatomical and physiological differences we see in the different parts of our bodies are, therefore, Grace manifesting Himself in different exquisite forms:

Our arms are Grace. Our eyes are Grace. Our hair is Grace. Our heads are Grace. Our facial expressions are Grace. Our emotions are also Grace. Every part of the human body—including its form and function—is Grace showing Himself in the forms we see and have become familiar with.

A direct implication of the great reveal is that the body does not only depend on food to live. Food does not sustain life; and cannot. Life sustains itself because life is Grace. Food maintains the biological processes, which are used by Grace to sustain the natural component of the human person. Only Grace can, and does, feed the spiritual component of the human person.

The biological processes we have widely studied and mastered in biology are only the superficial parts of the dynamics that sustain the human life. The human body depends 100% on Grace, who is Jesus Christ; because Grace is not only the life in our body parts. Grace is also our body parts, as well. All things originate from Grace.

From *1 Corinthians 6:13-15* it becomes clear, then, that Grace does not only reside inside human beings. The entire human biology and life's processes are constituents of Grace in the natural.

Our very own bodies are parts and parcel of Grace—that means full saturation with Grace. We are 100% grace—our body, our soul and our spirit. In other words, all of who we are is Grace who is Jesus Christ. Our body is Grace in the natural; and our soul and spirit are Grace in the supernatural.

So when scientists look at biological processes and components and conclude that they have it all figured out, it is nothing but a figure of their imaginations. They have not even scratched the surface.

What they have observed, they do not have full understanding of. No knowledge of anything in existence is complete without full knowledge and understanding of the Grace that makes that thing what it is. In essence, lack of the knowledge and understanding of God denies anyone full knowledge and understanding of any natural thing; because Grace is the only reason why anything exists and continues to exist.

Our biological books are filled with detailed descriptions of life processes that use different energy and material sources. But nothing in our science talks about the true substance of life and life processes—Grace. There is no life in anything unless Grace pre-exists in that thing. Grace starts all life; and Grace maintains all life and life processes.

No life exists inside anything without Grace first existing inside that thing. No vitality exists in anything on the earth and in the universe without Grace first existing in that thing. God builds biological life processes around Grace as the foundation and the powerhouse. In the same manner, through Grace, God also made and continues to enrich stars and all other entities around the universe with vitality.

Every single biological cell in all living organisms, intake Grace and uses Grace up the same way the biological cell uses up energy and other natural substances. If Grace is absent in biological processes, not only will the biological process not do anything; it will vanish altogether. Grace is the reason why anything exists. Without the pre-existence of Grace, nothing will exist anywhere.

Grace is the life that sustains all biological cells and processes, and keeps them in existence. In essence, Grace is biology. Grace is Chemistry and Physics. Grace is all knowledge. And Grace is everything that exists everywhere. The study of anything anywhere on the earth and in the universe is the study of Grace. No knowledge of anything is complete without the knowledge of how Grace works that thing.

Pick up any advanced biological textbook and see that each biological theory in the book breaks down at one point or another, forcing the experts that wrote the books to start ascribing god-like powers to biological particles and processes. That is the limitation of the human science. The compulsion of the human scientists to exclude God from public discussions has put scientists in that dilemma.

The world governments who are currently spending billions of dollars and man hours looking for the so-called mass-imparting particle, which was ironically dubbed the God-particle, are actually trying to detect Grace—the finest particle God has ever created, and the most abundant of everything in the world and the universe.

But no one can detect Grace through experimentation, because no one controls anything that is more powerful that they are. Every part of God has the power and potency of all of God, just like in a hologram. That is why God was able to create the entire universe from the invisible particle called Grace.

The Spirit is infinitely more powerful than all of nature! Grace can only be fathomed through faith in God, because Grace is the life-giving Spirit who is Jesus Christ. And the world and the entire universe are fully saturated with Grace—*"**who is over all and through all and in all.**"* *(Ephesians 4:6).*

Human science's estimation of the invisible universe is 96% of the entire universe. That unidentifiable and undetectable 96% of the universe represents the overflowing Grace which saturates the world and the universe completely.

Scientific knowledge is not separate from Bible knowledge. Scientific knowledge is a validation of the Bible knowledge, metered out to mankind as the humankind advances; to allow us to decipher deeper mysteries of God.

The 96% of the universe science could not see, but knows that is all around us, is made up of Grace and spiritual beings. And because the world scientists are stricken with unbelief, that undetectable 96% of the universe would forever remain elusive to human science.

God is not a thing that science can dissect and study. And since Grace is the Spirit of God, science's unbelief would never allow science to decipher the mystery of Grace. I know that many of you who are Christians or Jews, who believe strongly in science as separate from God, would have a hard time believing the revelations in this book. Yet, it is the uncompromising truth of God, given to me through revelations.

Your faith in God must be superior to your trust in the human science for you to be able to understand the mysteries of God revealed in the Bible. Conversely, any person's unreasonable dismissal of scientific information; makes that person ignorant of God's revelations through knowledge.

Ever wondered why there are four gospels, instead of one. That was not an act of man, but the grand design of Grace. God knows mankind's fixation on congruency and natural validation. The theologians—those experts who employ human science and empirical means to sign off on Bible records—would quickly agree on whatever appears consistent through the gospels, and cast doubts on the large majority each of the gospels exclusively recorded.

Most theologians quickly reach consensus on the gospels of Matthew, Mark and Luke; and widely differ on the gospel of John. Grace seeks to maintain the purity of faith. The biggest mysteries are deposited in the gospel of John, the apostle who was the most observant of the things that were taking place in the course of Christ's earthly ministry. In addition, John had a stronger spiritual connection with Jesus Christ and, His natural mother, the Virgin Mary. As a result, Apostle John has the most intimate connection to Grace—Love!

Next to Grace, in chaste and potency, is Love. And which apostle received a lion's share of Love from Jesus Christ? Apostle John, of course! John saw it all and lived through it all, constantly lurking at the corners even when he did not get the invite—like the time, after His resurrection, Jesus Christ summoned Peter, and John instantly followed them, causing Peter to protest to Jesus Christ to no avail.

Apostle John was the wholesome witness to the Life, the Light and the Way. That is why his gospel was the only one that started with the supernatural connection of God, His Son Jesus Christ, Grace and all creations. Jesus Christ, while He hung on the cross, personally handed His earthly mother over to

John minutes before He breathed His last:

"When Jesus saw his mother there, and the disciple whom he loved standing nearby, he said to her, "Woman, here is your son," [27] and to the disciple, "Here is your mother." From that time on, this disciple took her into his home." (John 19:26—27)

John was the only disciple of Jesus who had the fortitude to stand at the foot of the cross and who watched Christ's whole sufferings and insults.

So, while theologians, through their consensus, lead borderline believers to profess faith in Jesus Christ; only those who have unshakable faith receive the power to discern the mysteries God posited in the gospel of John, John's epistles and the Book of Revelation. The Book of Revelation was personally recorded by John while he was banished to a desolate island.

Even the church, every now and then, seeks the validation of modern science to authenticate its beliefs. That consists a clear and irrefutable indication that even the church has come to place the power of the mind over the power of the spirit. The human spirit does not seek to receive validations from the human mind. The human spirit responds only to the approval of the Spirit of God—who is Grace.

So, we must all embrace science through faith in

God; and not try to develop faith in God through our knowledge of science; or the validation of theologians. Scientific knowledge must confirm what you already know through faith in God.

Scientific knowledge does not lead to faith; it keeps you informed! Without the faith that comes from simply trusting the word of God, you will still lack discerning power. And without discernment, your faith will change with every scientific knowledge or theological argument *(Ephesians 4:14)*, minimizing your access to the surpassing power of Grace.

The Bible tells us that Grace comes from Jesus Christ and through Jesus Christ. That is why the church proudly recites the following passage from the Bible:

"May the <u>Grace of the Lord Jesus Christ</u>, and <u>the love of God</u>, and <u>the fellowship of the Holy Spirit</u> be with you all" *(2 Corinthians 13:14).*

Hence, the frequent associations: God is Love! Jesus Christ is Grace! And the Holy Spirit fellowships with all mankind! Most Christians who recite the above passage do not take to heart what they are really saying. Grace does not only come from Jesus Christ and through Jesus Christ. Jesus Christ is Grace and Grace is Jesus Christ! And for this reason, Jesus Christ declared to all mankind:

"Very truly I tell you, unless you eat the flesh of the

Son of Man and drink his blood, you have no life in you. ⁵⁴ Whoever eats my flesh and drinks my blood has eternal life, and I will raise them up at the last day. ⁵⁵ For my flesh is real food and my blood is real drink. ⁵⁶ Whoever eats my flesh and drinks my blood remains in me, and I in them." (John 6:53-56)

Then, He added:

*"Just as the living Father sent me **and I live because of the Father**, **so the one who feeds on me will live because of me**." (John 6:57)*

You heard it; and you heard it right! All human beings live because they feed on Grace who is Jesus Christ; just as Jesus Christ lives because He feeds on God the Father, who is also Grace.

And Grace is the invisible and uncontainable Holy Spirit of God who saturates the whole world and the entire universe. That makes the Holy Spirit one and the same with God the Father and God the Son—hence the Trinity. And since each of the Trinity is Grace, the Son, Jesus Christ, is also Love and brings fellowship to all mankind who cares for the Fellowship of the Spirit. This is why Jesus Christ declared in the gospels:

"I and the Father are one" (John 10:30).

Adding,

"I have much more to say to you, more than you can now bear. ¹³ But when he, the Spirit of truth, comes, he

will guide you into all the truth. He will not speak on his own; he will speak only what he hears, and he will tell you what is yet to come. ¹⁴ **He will glorify me because it is from me that he will receive what he will make known to you.** ¹⁵ *All that belongs to the Father is mine. That is why I said the Spirit will receive from me what he will make known to you."* (John 16:12-15)

If you are appalled by this revelation that all human beings must assimilate Grace who is Jesus Christ to live, you are not alone. The disciples who heard Jesus Christ make this proclamation more than two thousand years ago had a similar reaction. However, their skepticism did not spare them the wrath of God. And our skepticism today would not save us from it either. The Bible says:

"On hearing it, <u>many of his disciples</u> said, "This is a hard teaching. Who can accept it?"" (John 6:60)

And to that, Jesus Christ said:

"Does this offend you? ⁶² Then what if you see the Son of Man ascend to where he was before! ⁶³ <u>The Spirit gives life; the flesh counts for nothing. The words I have spoken to you—they are full of the Spirit and life</u>. ⁶⁴ Yet there are some of you who do not believe" (John 6:61-64).

He then added,

"This is why I told you that no one can come to me unless the Father has enabled them." (John 6:61-65)

The Bible says that most of His disciples stopped following Him. Here is that passage:

"From this time many of his disciples turned back and no longer followed him." (John 6:66)

666! Does this number ring a bell? You got it! It is this same Apostle John who received the eternal word of God about the end of all things in the Book of Revelation that posited this information in *John 6:66— From this time many of his disciples turned back and no longer followed him.* The active product of the human 'will' that caused these disciples to walk away and followed Jesus Christ no more some two thousand years ago, is the same thing the Book of Revelation says will bring the end of the world.

The falling away that was prophesied in the Bible has already begun, and increases every passing day—driven by the world's exclusive dependence on human science. Science is the false prophet who will mesmerize the world with fascinating things, and get the world to follow the beast—the antichrist. Look at the following passage from the Bible and see that science and government are the fulfillment of everything in the passage:

"Don't let anyone deceive you in any way, for that day will not come until the <u>rebellion occurs</u> and the man of lawlessness is revealed, the man doomed to destruction. ⁴ <u>He will oppose and will exalt himself over everything that is called God or is worshiped</u>, so that he sets

himself up in God's temple, <u>proclaiming himself to be God</u>.

Don't you remember that when I was with you I used to tell you these things? ⁶ And now you know what is holding him back, so that he may be revealed at the proper time. ⁷ For the secret power of lawlessness is already at work; but the one who now holds it back will continue to do so till he is taken out of the way. ⁸ And then the lawless one will be revealed, **whom the Lord Jesus will overthrow** *with the breath of his mouth and destroy by the splendor of his coming. ⁹ The coming of the lawless one will be in accordance with how Satan works. He will use all sorts of displays of power through signs and wonders that serve the lie, ¹⁰ and all the ways that wickedness deceives those who are perishing. They perish because they refused to love the truth and so be saved. ¹¹ For this reason God sends them a powerful delusion so that they will believe the lie ¹² and so that all will be condemned who have not believed the truth but have delighted in wickedness."* (2 Thessalonians 2:3-12)

While mankind looks at Bible prophecies from a natural perspective, relative to man; God looks at everything on a colossal scale, because God is infinitely larger than all of His creations put together. And science is a spirit of colossal size. That is why science's reach touches the whole earth. The same also applies to 'government'—the dominion of the world; with individual governments as images of the beast.

From John 6:66 and the Scripture verses preceding it, it can clearly be seen that the risk of losing salvation is not to those who have not had the opportunity to hear the gospel; as it is to those who have received the gospel, but through their own reasoning and informed decision, walked away from the righteous path to salvation—just like the disciples who walked away in John 6:66.

This brings us to the two criminals who were crucified alongside Jesus Christ. Those who had not heard the gospel would hear it and readily accept it up until the Last Day—just like the criminal that rebuked the other, and asked Jesus to remember him when He gets into His kingdom. And also like that criminal, they will instantly be granted salvation by Jesus Christ.

And those who had received the gospel, and had for a while enjoyed the favor of Grace, but later walked away from Grace to chase the thrills of their own minds; those people had taken the number of the beast and are serving the beast. And like the criminal that hauled insults at Jesus Christ as he hung from the cross, this group of human beings will perish with the devil in the lake of eternal fire.

Christians must worry about their fellow Christians missing out on heaven, more than they are concerned about people of other faiths missing out on heaven; because those who have already embrace the faith, who however goes in and out of the faith, are already much farther down the path of destruction than those who have not heard the good news of the kingdom.

All church organizations; and all Christian denominations; and all believers of God and His Christ; must teach those who have already come into the kingdom of God how to remain in the kingdom. This is a matter of first priority. Operating within true kingdom economic principles then becomes a matter of urgency to all the churches and the kingdom citizens.

It is not the opinions of the theologians or the scientific arguments that saves anyone. The academics will do what academics always do—train the mind to decide everything for itself. It is having unwavering faith in the simple word of God that positions anyone to discern the mysteries of God that are hidden in God's words.

The disciples of Jesus Christ, who, more than two thousand years ago, heard Jesus say that mankind must assimilate "Grace who is Jesus Christ" to continue to live, were torn between their spirits and their minds.

They wanted to believe Jesus Christ, but their minds overrode their spirits, and they left the path of righteousness, and went back to their worldly ways. They could not find any justification in the real world to substantiate what Jesus Christ just proclaimed to them. Dismayed and confused, they walked away and no longer followed Him.

We all spend so many years in schools training our minds; but no time at all, tending to the needs of our spirits. How then can the spirit compete? Even the knowledge, which God deposited on the pages of the Bible to enrich our spirits, has been doctored by the world to subvert the spirit, and further elevate the mind.

The devil has not only been attacking from outside the church. The devil had penetrated the church and adulterated the purity of the church. And this has taken the minds of those who genuinely care about the word of God off of what the true war really is.

The war between good and evil is not "we" versus "they". The war between good and evil is raging inside each and every one of us. Every single human being that has walked on the face of the earth has partaken of this battle. The war is between the spirit inside each human being and the mind of that human being. So, it is a personal way. This is why salvation is decided by Grace on an individual basis.

Through history, believers have been banding together and formulating doctrines, hoping their doctrines would give them advantage over the devil; validate their approach to the commands of God; and assure them of the favors of Grace. But because they have completely failed to identify the real battleground, all their doctrines still fall short; and members of their congregation, including their leadership, have continued to incur great loses in the same magnitude and frequency as the rest of the world.

Grace rewards or punishes automatically! Grace acts at every instance of each human person's action: Grace instantly rewards each good deed; and each instance of obedience to the commands of God; and each instance of repentance and reparation. And Grace instantly allows for the punishment of each wicked deed and each instance of disobedience to God's command.

The dynamics of Grace and the activated evil is best illustrated by the effect of temperature. Throughout human existence, mankind has felt both heat and cold. Then came science, which announced to us that cold is really not a separate thing: but simply, the absence of heat. That both the heat and the cold operate along a single line of temperature gradient. Yet, reality tells us that cold is different from heat because they have two different effect on the rest of nature.

Same can be said of light and darkness. Science tells us that darkness is nothing but absence of light. Yet God tells us that there are no two things that are as different as light and darkness. And real life tells us that both light and darkness affect nature differently. So, who do we believe, God or science? That one effectively obliterates the other does not mean that that which is eliminated is not an entity by itself. God says that it is. Then, it is!

That light overwhelms darkness does not make darkness "nothing". And that heat removes cold does not make cold nothing. Both cold and darkness are real creations of God—both independent and real. That their properties are mirror images of their opposites does not change what they are created to be, and what they are created to accomplish.

This is the same dynamics for Grace and the activated evil. Grace drives everything that God has ever created. But on the opposite side of Grace is evil—inactive evil (the product of the evil spirits), and activated evil (the product of human disobedience of God and human wickedness).

When God reduces Grace in anything—natural or spiritual—God allows evil to automatically take up the space Grace has vacated. And the evil that comes in immediately goes to work against the life of the person it comes into. That is why the human person must always strive to be full of Grace. There is no middle position when it comes to Grace and evil. Together, they make up the fullness of spirit in any living being—spiritual or natural.

Just as light drives away darkness and heat drives away cold, Grace drives away evil. That is why believers are taught to ask for Grace, and more Grace. The continual balancing of Grace and the evil particles within each human person is the war between good and evil that is continually fought within each and every one of us.

The venue for that balancing and counterbalancing is the human mind—Grace on one side and the seed of the devil on the other side. And the products of that war is either more Grace produced from within the individual (rivers of living water), if one triumphs; or activated evil produced from the seed of the devil, if one succumbs to the desires of his mind.

To succeed in life, we must each understand the real venue for the battle between good and evil. Otherwise, the devil will take us all out, one by one; even as we observantly watch and thread cautiously. The war between good and evil is not on a worldly stage, even though it appears that way most of the time.

What we see on worldly stage is only the result of a war that had already been lost by the individuals perpetrating the wicked acts we see every day around us. They have each lost the real war before they showed up on the world stage to showcase the devastation of their loss. The battle rages on inside each and every human person, including our children who are the most vulnerable to the devil's insidious schemes.

The devil picks his human victims one person at a time, starting with the most vulnerable, and going up. That is why Eve fell before Adam followed. Once the devil took Eve out, Adam did not stand a chance. He went down too, without a fight or any resistance—Adam simple accepted the fruit from the forbidden tree from his wife and ate it.

While Adam sat there watching his wife go back and forth with the devil, he was under attack as well, by the devil; but he did not know it. The listener is as vulnerable as the person engaging in an argument. The listener listens to both sides, and takes in their different points of view. The things he hears from both sides make impressions on his mind.

And by the time the argument is over, he has decided one way or another. Or he is forced to become a party in the argument by the power of what he has heard. Either way, he can no longer sit on the sideline. He has been permanently affected by the argument. The devil was trying to get to Adam by engaging Eve, and his sinister plan worked, and they both fell.

That is why the Bible warns sternly:

"Keep reminding God's people of these things. <u>Warn them before God against quarreling about words; it is of no value, and only ruins those who listen.</u>" *(2 Timothy 2:14)*

The church has not only been focusing on the wrong beast, but has actually greatly helped the devil. Instead of the church working to minimize the influence of the human mind over the human spirit, the church has been greatly elevating the human mind over the human spirit, in the name of civilization. And by so doing, the church has helped make the enemy more powerful.

Human beings can educate the human mind, and have been doing so for centuries. But only God invigorates and reinvigorates the human spirit. But this reinvigoration of the human spirit is only possible when the person whose spirit needs reinvigorating constantly lends their spirit to Grace.

Nowadays, that war between each man's spirit and his own mind has gotten a whole lot worse because the world—being truly of the devil—has exclusively trained the mind to resist everything Spirit, including the spirit inside every man, every woman and every child.

Through science and astronomy, and the extremely fast news and information dissemination media, the world has successfully focused every human person entirely on dominating their spirit with their mind.

The human person now relies on his mind, exclusively, to come up with solutions to his every problem. In allowing his mind to subdue and dominate his spirit, the human person has turned his life over to the devil, who has great influence over the mind of mankind; in effect turning his back on God. Majority of mankind no longer knows how to reach their own spirits; because the world has caused all to abandon their spirits, by focusing exclusively on the education of their minds.

A great majority of the world, who enjoy the fruits of their minds so much, has now, not only abandoned religion, but has been bashing religion and God with all their energies, and all their resources. And many in the church profess Jesus Christ not because they are in touch with their spirits, or working to be in touch with their spirits, but because they know it is a safe position to be in. Because of all these, God declare through Prophet Jeremiah:

"My people have committed two sins: They have forsaken <u>me, the spring of living water</u>, and have dug their own cisterns, broken cisterns that cannot hold water." (Jeremiah 2:13)

And Prophet Jeremiah proclaimed:

"Lord, you are the hope of Israel; all who forsake you will be put to shame. Those who turn away from you will be written in the dust because they have forsaken the Lord, the spring of living water." (Jeremiah 17:13)

God is never threatened by any human situation, because God is supremely confident of His control over all His creations. His Grace saturates all of His creations—empowering, monitoring, replenishing, restoring, repairing, substituting, rebuking, punishing, correcting, and controlling all of them in real time. His purpose will always stand!

The Bible declares that God does not despise man because mankind is but a breath which is easily withdrawn by God. God is gravely concerned that mankind has chosen a disastrous end for itself. God—out of His great love for mankind—does not want any human being to perish.

God is also concerned that the drumbeaters for this monstrous atrocity of smoldering the human spirit by the human mind are, by the day, drawing hundreds of thousands of people into that horrific destiny that is sure to come.

Did God not prophesy this from long time ago: that the world—like Eve—would go the way of the mind and forsake the Spirit? Did God not declare to mankind that this world will come to an abrupt end, without a warning, just like in the great flood of Noah; and like in the fiery destruction of Sodom and Gomorrah?

God's prophecies are fulfilled, not so God will show human beings that God could predict the end right from the very beginning of time; but because God knows that mankind would not change its direction, because mankind is hooked on the pursuit of personal desires and pleasures—which are the lures of the devil—the products of the human mind.

Therefore, the world will abruptly be destroyed by God, not to prove God right, by fulfilling the prophecies of God; but because the world continues to bring destruction on itself by activating evil particles that continue to accumulate and rise to suffocating levels in our world.

God continues to work our world and our universe through Grace. The entire universe and the earth and everything in them are saturated with Grace. The voids and all empty spaces that constitute the majority of the universe and the earth are saturated with Grace. And Grace comes from the Spirit of God through Jesus Christ the Son of God.

When Lucifer and some angels of God turned evil, their new form of existence began to generate a new particle and release the new particle into the world and the universe. And this new development was not out of the devil's power or the power of his evil companions. It was a design by God, to segregate the evil "fruit" of the devil from God's "fruit" of love, which is Grace. Hence, "inactive evil" particles were born.

God designed all living beings, not only to assimilate Grace and use Grace in sustaining their lives and life functions, but also to generate Grace by their good deeds and their obedience to Jesus Christ, and their love of God, and their reverence to the Holy Spirit. This is why Jesus Christ told us in the gospel:

"Whoever believes in me, as Scripture has said, <u>rivers of living water will flow from within them.</u>" *(John 7:3)*

What are these *"rivers of living water"* that *"will flow from within the believers of Jesus Christ?"* The living water is Grace. And Grace is Jesus Christ. Because God had created each and every human being and placed a portal inside each person to generate Grace as we obey God and do good deeds to one another, when anyone truly believes in Jesus Christ, the person's faith in Christ automatically activates that portal which God has placed inside the person.

And from this portal comes abundance of Grace. Every aspect of our holy lives—thoughts and deeds—opens up a new source somewhere in the portal, to generate Grace. And each new source turns into a river of Grace. And together, they make up the rivers of Grace which Jesus Christ was referring to in His declaration in the gospel. So all human beings are not only designed to consume Grace, they are equipped to generate Grace through their individual faith in God, to enrich their lives and empower themselves.

And the Grace that originates from within each person can also be imparted to others to help them start their journey of faith. This is the impartation of the Spirit that is described in the Bible. Grace is Spirit because Grace is Jesus Christ. And Grace is the Living Water. This is why Jesus Christ said to the Samaritan woman at the well:

"If you knew the gift of God and who it is that asks you for a drink, you would have asked him and he would have given you <u>living water</u>." (John 4:10)

And when God gives Grace, God gives Grace abundantly. This is why the Bible says that God lavished Grace on us:

*"Blessed be the God and Father of our Lord Jesus Christ, who has blessed us in Christ with every spiritual blessing in the heavenly places, ⁴ even as he chose us in him before the foundation of the world, that we should be holy and blameless before him. ⁵ He destined us in love to be his sons through Jesus Christ, according to the purpose of his will, ⁶ to the <u>praise of his glorious Grace which he freely bestowed on us in the Beloved</u>. ⁷ <u>In him</u> we have redemption through his blood, the forgiveness of our trespasses, **according to the***

riches of his Grace ⁸ which he lavished upon us. *⁹ For he has made known to us in all wisdom and insight the mystery of his will, <u>according to his purpose which he set forth in Christ</u> ¹⁰ <u>as a plan for the fullness of time</u>, to <u>unite all things in him</u>, things in heaven and things on earth.*

¹¹ In him, according to the purpose of him who accomplishes all things according to the counsel of his will, ¹² we who first hoped in Christ have been destined and appointed to live for the praise of his glory. ¹³ In him you also, who have heard the word of truth, the gospel of your salvation, and have believed in him, were sealed with the promised Holy Spirit, ¹⁴ which is the guarantee of our inheritance until we acquire possession of it, to the praise of his glory." (Ephesians 1:3-14)

The same Jesus Christ who the preceding passage so clearly describes as the shield of all mankind and our hope for salvation; who has given us great knowledge and has revealed His mysteries to us to bolster our hope in the things to come while in the meantime living off of His Grace; will be our final destination.

His Grace began all our lives when we were, each, created by God. Currently, His Grace fuels us, nourishes our bodies, enriches our minds, heals our wounds, and protects us all from activated evil. Finally, His Grace would become everything to us, through all eternity. Here is the Bible telling you about this final leg of our journey—those who did not let their minds mislead them:

"For the Lamb at the center of the throne will be their shepherd; 'he will lead them to <u>springs of living water</u>.' 'And God will wipe away every tear from their eyes.'"
(Revelation 7:17)

When a priest of God raises a piece of bread in faith, and blesses it in the name of the Father, the Son and the Holy Spirit, he is inviting Jesus Christ to saturate that piece of bread with Himself—the abundance of Grace! When the priest does the same thing with a cup of wine, the wine also gets saturated with Grace—that life-giving Spirit who is Jesus Christ.

That is how the piece of bread super-scientifically transforms into the body of Jesus Christ; and the wine, super-scientifically, transforms into the blood of Jesus Christ. It is real science, but one that is above and beyond our human science; and not a symbolism, as even some Christians believe!

It is faith—through our spirit—that activates Grace and gets Grace to work for us; and not the brilliance of our mind that activates Grace. Grace simply does not care about the brilliance of anyone's mind. Grace frustrates the pomposity of anyone's mind, because Grace encourages humility.

A priest also gets Grace to saturate every other blessed article the priest distributes to the believers. Grace responds to every expression of faith in Jesus Christ and in God.

The Grace who is Jesus Christ saturates the whole world and the universe completely so that through faith in God, a lion's share of Grace is added to anyone who believes. That is why the Bible tells us to seek so we may find; and to ask so we may receive; and to knock so it may be opened to us. It is all done to us by Grace.

Grace is constantly all around us; going into us, and passing through us. Grace connects all of us, in real time, to God. And through Grace, God monitors and controls all of us and everything else that exist on earth and in the universe.

Grace automates everything in existence. Grace automatically rewards good; and Grace automatically punishes evil. And Grace connects all of us to one another. Only our faith in God and His Christ unlocks Grace. Grace is our assurance to salvation and to life!

If your education has made you question the Bible and faith in God, you are already on a slippery slope and have to rush to regain control of your life so that the Grace of God would inundate your life and direct your path in life.

If your job requires you to shun even the slightest mention of God and religion, you are in the snare of the devil, and have already lost the power to save yourself from the destructive powers of the devil. But the good news is that inside of you still resides the portal for the living water—Grace.

God never shuts this portal down. All anyone needs to do to activate this portal is to turn their face to God and ask for His forgiveness and His mercy, and the reversal would begin for them. God's mercy is always available to everyone who asks for it. And the only condition to receiving God's mercy is to genuinely seek it and ask for it.

If your knowledge in science and related fields of study has caused you to discredit God in anyway, even so slightly, you are fast losing the Grace of God. You are activating evil at an alarming rate and saturating your body and your life with activated evil—all on your own.

In addition to activating evil particles all around you, Grace is leaving your body and your life at an alarming speed, making your body faculties and your entire life accessible to the devil; and susceptible to rot and decay caused by activated evil.

Your person and your life are not only at the mercy of the evil you personally activate, but also the evil activated by those you encourage in their evil ways; or who you support and defend. The evil they activate affects not only their respective lives;

it also affects your life as well.

But when you openly condemn all evil, you retain your supernatural buffer against activated evil, including the evil activated randomly by your occasional errors of judgement, because the Grace that is generated from within you—as "the rivers of living water" through your obedience of Jesus Christ, your love of God, and your good deeds to other people—increases to great levels to protect you from activated evil. This is what the Book of Job refers to as God building a hedge around a believer, the believer's family and everything the believer has.

You are your own best advocate with respect to Grace; and you are your own worst enemy with respect to activated evil. You choose the direction you want your life to go. And that is what you receive.

The choices that you make, determines the course of your life. This is an automatic thing. God intervenes only if He chooses and when He chooses. Your own destiny is in your own hands. Therefore, release the "rivers of living water" from within you and bask in the Grace of God and the richness of life. All it takes is your unwavering faith in Jesus Christ.

Ladies and gentlemen! One with God is still majority; and will continue to be majority, because of the abundance of Grace available to the person!

When you trust God and simply submit to the commands of Jesus Christ in all areas of your life, God, through personal signs and wonders, continually communicates to you His approval of the way you live your life. This is God's way of encouraging you to never waver in your commitment to Jesus Christ.

Ever wondered how the young Joseph managed to remain faithful to God throughout his ordeal in Egypt? Ever wondered how Jacob remained level-headed with his father-in-law Laban even after much mistreatment? Ever wondered why Ruth could not leave her mother-in-law behind? They looked for signs of God's approval in their daily activities and received signs that made them continue to believe.

So, they remained humble as life mistreated them, because they knew that God was with them through it all. God never fails to give anyone who remains loyal to Him personal signs and wonders for reassurance. God remains the wings beneath your wings when you place all your trust in Him and believe in all His commands.

CHAPTER TWO

The Threshold of Life!

Life! Does life have a threshold? The Bible declared that life is in the blood. The Bible also declared that life comes from Jesus Christ and through Jesus Christ. Through medical science, we have understood that certain level of blood is required to be in the body for the body to continue to live. Therefore, the life in living things has a threshold. Thresholds like in the electrical energy inside a battery or through different electrical wire gauges? You got it!

What then is the threshold of life? The threshold of life is, the level of Grace required for a living being to remain alive and continues to perform life functions. Different living beings have different thresholds for Grace, to become, and remain, the living being that they are. God assigned various threshold of Grace to different beings, and apportioned Grace accordingly to each class of living being to achieve His intended purpose.

God then hinged life at the various thresholds, so that extra Grace will be required to unhinge life, before life could be drained completely from a living body to allow death to set in. Life does not just leave a body because of trauma to the body, or the ravages of a disease. Grace has to first unhinge life before life can leave a biological body.

Like electricity in a light bulb, right before the light bulb dies, there is a surge in illumination. And this surge cripples the filament, and the light flickers out. Similarly, right before life expires (as in death), life inside a person surges, making one appear to be recovering from whatever is draining life out of the body. But soon after, life leaves the body of that living being for the last time.

This surge is caused by the source which powers the body—and not from within the body. In the case of the light bulb, it is the electrical power—external to the light bulb—that creates the surge. And in a dying person, the surge of life right before death is caused by Grace, to unhinge life and permit life to leave the body permanently. And Grace is God the Father. Grace is God the Son and Grace is God the Holy Spirit. That is why the Bible says that God gives life, and God takes life away! And the Bible refers to death as "God gathering a person up."

In *Genesis 1:14*, God said that the universe was to serve as signs for mankind about the mysteries of God, and it does in more ways than mankind realizes. Good observations of the natural things around us allow us to better understand the spiritual things God tells us in the Bible. Here are some examples of the universe serving as signs to mankind:

God started the creation of the heavens and the earth inside the watery depth in the hollow of God's hand *(Isaiah 40:12, Proverbs 8:22-31, Genesis 1:2, Psalm 95:4)*. And in the natural, God starts the birth of life in a watery medium inside a woman's womb. Not just the birth of the natural human person, but the birth of all natural living beings—including the Son of Man, who took the form of His brothers so He could save His brothers from Satan's bondage.

God pioneered the creation of natural life in the open space at the Garden of Eden. And after the fall of mankind, God, by way of Grace, moved the creation of natural life into the womb of the woman to continue His eternal purpose, which was interrupted by sin.

The same way Adam and Eve were created in the open at the Garden of Eden through Grace, without the knowledge of good and evil; through Grace also, a human child is also created inside a woman's womb without the knowledge of God and evil.

And the same way Adam and Eve lived without the

knowledge of good and evil until their disobedience of God at the Garden of Eden, a human child is blameless from sin until the child could choose right and reject wrong—that is, until the child reaches the age of consent.

The same way Adam and Eve experienced shame, guilt, remorse and other wild human emotions when they disobeyed God and activated evil on the earth, the growing human children experience shame, guilt and other negative emotions when sexual awareness and desire first manifest in them; and their parents take notice and try to offer some advice or caution. And again, like Adam and Eve, once they overcome these negative emotions, they join the rest of humanity in its sexual pervasion and its insensitivity to the commands of God.

The way it takes specialty natural knowledge to detect and understand natural mysteries, is the same way it takes specialty spiritual knowledge (discernment) to detect and understand the spiritual things God is saying to us in the Bible.

The way it takes specialty knowledge of science to understand that the visible light is an extension of the invisible light, is the same way it takes specialty spiritual knowledge to understand that the natural world is an extension of the supernatural world.

The natural knowledge of different forms of energies, their interchangeability, their consumption and dissipation, and their effects on natural processes lends itself to the spiritual discernment of Grace in relation to the Trinity, the propagation of Grace through everything across the earth and the universe, the consumption of Grace in all dynamic systems which is comprised of the entire earth and the universe, and the perfect synergies which Grace produces in all the creations of God.

The earth's oscillatory movement—23 degrees north to 23 degrees south, for a total of 46 degrees displacement; serves to equilibrate sun's energy between the northern and the southern hemispheres, creating the different seasons of the year and maintaining life's equilibriums on the earth.

Grace compels the earth to tilt 23 degrees north and 23 degrees south to regulate life on the earth and maintain the desired balance! Correspondingly, the quality of life within human beings is controlled by 23 pairs of chromosomes. One chromosome in each of the 23 pairs comes from the female and the other in each of the 23 pairs of chromosomes comes from the male—just as in the 23 degrees to the north and 23 degrees to the south oscillatory movement of the earth. Both situations are predestined, accomplished and controlled by Grace. Grace controls everything in the world and in the universe!

A woman's menstrual cycle is 28 days, which corresponds to the moon cycle. From one appearance of the moon to another is twenty eight days—and so is the woman's menstrual cycle, which does not only control her fertility but also her health and her quality of life; in addition to the effect her menstrual cycle has on her marriage. The solar cycles mimic the cycles of life. And the cycle of life centers on humanity. The human person is the pivotal interest in God's creation of the natural world. He created us in His image and His likeness.

A woman's menstrual cycle is not a mere biological event. It has spiritual embodiment that surpasses its biological significance. That is why the Bible says that a woman will be saved through childbearing, if she continues in faith, love and obedience to God. Here is that passage from the Bible:

"But women will be saved through childbearing—if they continue in faith, love and holiness with propriety: (1 Timothy 2:15)

The woman's menstrual flow is spiritually tied to the cleansing power that legitimizes marital sex in the eyes of God— *"Your desire will be for your husband"*—*Genesis 3:16*. Her flow purges her spiritual impurities and those of her marriage. This is why all other kinds of sexual relations are forbidden by God. While sex within marriage helps to fulfill God's purpose of bringing holy children into the world, all other forms of sex simply pleasure human beings. A woman who is undergoing her monthly flow is deemed spiritually unclean. Here are the Scriptures:

"When a woman has her regular flow of blood, <u>the impurity of her monthly period</u> will last seven days, and anyone who touches her will be unclean till evening." (Leviticus 15:19)

"If a man has sexual relations with her and her monthly flow touches him, he will be unclean for seven days; any bed he lies on will be unclean." (Leviticus 15:24)

"If a man has sexual relations with a woman during her monthly period, he has exposed the source of her flow, and she has also uncovered it. Both of them are to be cut off from their people." (Leviticus 20:18)

To make everyone realize the seriousness of these commands, and many others, which God gave to His people, God commanded:

"Keep all my decrees and laws and follow them, so that the land where I am bringing you to live may not vomit you out. [23] <u>You must not live according to the customs of the nations I am going to drive out before you. Because they did all these things, I abhorred them.</u> [24] But I said to you, "You will possess their land; I will give it to you as an inheritance, a land flowing with

milk and honey." I am the LORD your God, who has set you apart from the nations." *(Leviticus 20:22-24)*

A woman who is going through her monthly period is going through spiritual cleansing—cleansing not only on her behalf, but also on the behalf of her marriage—if and when she is married. A woman comes into the life of her husband with this special gift to protect the two of them from the sinfulness of sex.

Since the natural provides us with evidence of the supernatural, it is safe to conclude that menstruation in the human female did not start until after the fall of mankind from Grace. A look at the developing human child also shows that girls do not start to menstruate at birth, but rather at puberty. Adam and Eve's rebellion against God marked their "puberty" since that was the point of their sexual awareness and their sexual desire—the desire which God pointedly warned Eve against in *Genesis 3:16*.

Childbearing through the natural process, as we know today, did not enter the human psyche until after the fall. It was because of the fall that God announced childbearing through the natural process, telling the woman that it was her who will be bringing children into the world, and that she would be doing that in pain—because of her disobedience of God.

Animals have chromosomes, but only man's chromosomal number aligns with the oscillatory limits of the earth in relation to the sun—earth's source of light and energy. And the sex chromosomes are designated as X and Y. For the woman they are XX and for the man they are XY. And to every baby born, the woman contributes the X, while the man contributes either the X or the Y. The Y chromosome determines what sex of a baby.

Therefore whether a baby is a boy or girl depends on the component of the sex chromosome that comes from the man. If the man's component is an X, a girl is born; and if it is a Y, a boy is born. Coincidence! Not at all! That is God's design at its best.

Everything God uttered from His mouth had already been designed, built and deployed. We humans are simply too slow to catch on.

It is God who decides, through Grace, the sex and destiny of every human being born into this world. The human genes do not make determinations as to what traits a human person receives. The human genes are the end products of the decisions made by Grace ahead of the spiritual conception of a human person.

The genes only expresses what had already been designed, established, completed, and maintained in them by Grace. Grace remained inside the Genes for the genes to hold the biological expressions imprinted on them.

If Grace is removed from anything, that thing will not only die; that thing will disappear altogether. Grace is not only life. Grace is existence, period.

Grace remains in everything—living or dead. When anything dies, Grace remains in them—below the threshold required for life. The remnant Grace is what allows the dead body to remain in existence and continue to be visible. Otherwise the dead body will disappear completely.

The genes are expressions of character traits; and not the determinants of character traits as science claims. Genes are incapable of determining anything.

Only God, through Grace, determines everything. Every gene has Grace saturating it completely, unbeknown to the scientist that ascribes "magical" powers to the genes.

Natural methods can neither detect Grace, nor could they separate Grace from anything in nature. Only spiritual submission to God allows anyone to experience the presence of Grace.

Grace must first exist inside of anything for that thing to have an existence and a form; characteristics and functionality. God created the world and the universe, and everything in them, from Grace, through Grace, by Grace and for Grace; because Grace is God the Father, God the Son and God the Holy Spirit.

That science's manipulations of the natural biological components of the living body result in expected outcomes do not mean that these natural biological components of the living body are responsible for the results. Grace is intrinsic in all natural and supernatural particles and components. Grace is the substance of everything!

It is the unseen, the undetectable, the immeasurable and the indomitable Grace—which saturates these natural biological components—that produce the various results we see at the end of our scientific manipulations. Grace is the supreme ingredient of life and vitality in everything. So, excuse me for qualifying Grace as "which" in the preceding declaration.

Science does not produce anything; in much the same way science did not create the world and the universe and everything in them. Grace created the world and the universe and everything in them; and Grace produces everything we obtain through our scientific manipulations!

That is the perfection of Grace. And this perfection produces repeatable outcomes in all natural orders when approached in consistent manners. The perfection of Grace guarantees that the result we obtain remains the same time after time.

We are all spirits, expertly packaged in organic materials for adaptation in a natural world. Therefore, the human genes are all organic packaging for spiritual traits assigned to each human person through Grace and by Grace—for the glory of God.

When scientists gloat over their discovery of a gene that control this and a gene that control that, they have only uncovered a packaging. They can neither detect the content, nor could they manage the content, because the content of that packaging that is the gene is Spirit. And man cannot handle or manipulate God!

Yet, we hear scientists brag about having mapped out the entire human genome, as if that entitles them to go ahead and start sculpting living and operable human beings. That is a disaster waiting to happen.

Their achievements are all superficial. Mankind does not have the authority to touch life, let alone decide life. Life is holy!

To human beings, everything leads to money: If we can decipher it and make sense through it, we commercialize it; or worse still, divert it to a dubious use, to gain bragging rights and attract more wealth.

When a wife and husband do their business to have a baby, they are simply lending their hands in fulfilling God's Will in the emergence of a baby. Nevertheless, God ensures that whether the baby is a boy or a girl must be decided from the man's contribution to that natural chemistry.

The human body is filled with divine coding to ensure that the will of God always prevails over the meddling of the devil and human beings. That is the supremacy of God.

A man is not only anatomically and physiologically different from a woman. He is wired differently, spiritually. His eternal duties and responsibilities are also different from those of the woman.

Before God said anything to human beings about His work and His purpose, God had already established these distinctions and safely tucked them away inside our bodies. God then provided clues about these distinctions in the sixty six Books called the Bible. How much of these distinctions we discover scientifically, or spiritually, depends on our obedience to God.

God made sure that the distinctions between men and women are complimentary to one another. That is why God promptly married Eve to Adam so that both of them uniting made them whole again:

they became one body and one spirit *(Genesis 2:24 & Matthew 19:6).*

Therefore, the shedding of the blood from the wife's menstrual cycle cleanses both her spiritual impurities and her husband's spiritual impurities, if they remained faithful to each other, and obey God fully.

And the spiritual impurities that come from sex are more pervasive because, as the Bible says, sex is the only moral filth that enters the body and violates it. Here is that passage from the Bible:

"You say, "Food for the stomach and the stomach for food, and God will destroy them both." **The body, however, is not meant for sexual immorality but for the Lord, and the Lord for the body**. *[14] By his power God raised the Lord from the dead, and he will raise us also. [15]* **Do you not know that your bodies are members of Christ himself?** *Shall I then take the members of Christ and unite them with a prostitute? Never! [16] Do you not know that he who unites himself with a prostitute is one with her in body? For it is said, "The two will become one flesh." [17] But whoever is united with the Lord is one with him in spirit.*

[18] Flee from sexual immorality. **All other sins a person commits are outside the body, but whoever sins sexually, sins against their own body**. *[19] Do you not know that your bodies are temples of the Holy Spirit, who is in you, whom you have received from God? You are not your own; [20] you were*

bought at a price. Therefore honor God with your bodies." *(1 Corinthians 6:13-20)*

The marriage of a man and a woman in a holy union has become their entire lives, because everything they do, they do as a husband and a wife—the two sides of the same holy union.

Some animals have ovulation cycles like humans, but only that of the human female aligns with the cycle of the moon—another supernatural sign to human beings that God is continually in charge. And the moon does not have its own light but reflects the light of the sun.

God created the woman as *"a suitable helper" (Genesis 2:18-24)* for the man in just the same way God created the moon to suitably help the sun continue to cast its light on sections of the earth that are out of the reach of direct sunlight.

We can, therefore, see that a suitable helper is not a slave to the one she is helping. The helper and the one she is helping are mutually complimentary in their joint existence and their joint derived benefits. A suitable helper helps the one she is assisting to tap into vital resources beyond his usual reach; while at the same time, drawing her support and vitality from the one she is helping.

A wife is a favor and an inheritance from God, brought into the husband's life to help the man reach vital areas usually inaccessible to him, for the mutual benefits of the two of them; and the family they raise together.

Only the females of primates—the group of animals classified by evolutionists with human beings—shed their endometrium at the end of their ovulation cycle. It is interesting to know that all primates are found within the tropics: within 23 degrees north of equator and 23 degrees south of equator. Here is an internet reference to that claim:

"With the exception of humans, who inhabit every continent, most primates live in tropical or subtropical regions of the Americas, Africa and Asia."
(http://en.wikipedia.org/wiki/Primate)

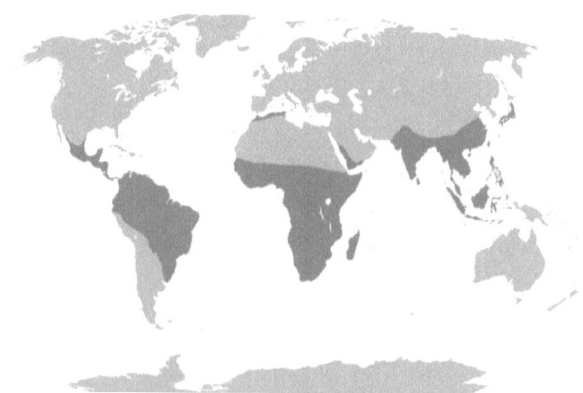

Figure 2: The darker shades on the map show where most primate populations are found throughout the world.

CHAPTER THREE

There Are More Particles Than We Know

There are numerous other forms of matter and energy (particles, waves, quarks, etc.) than so far discovered by man. And God continues to create new forms. Anything and everything discovered by mankind so far is a gift from God. Mankind cannot decipher any mystery God does not want mankind to decipher. Human beings are not independent beings. We are discrete but not independent.

Our existence is made possible by the Grace of God who is the very life that dwells inside each and every one of us—the very power that activates and fuels <u>everything</u> that we do. The Spirit of God that lives inside of us takes the signal from our will and accomplishes whatever it is that our will has chosen to accomplish. The only power any human being has is the power to decide one way or the other, in all the issues in our lives. That is why the Bible says:

"'For in him we live and move and have our being.' As some of your own poets have said, 'We are his

offspring." (Acts 17:28)

"There is one body and one Spirit, just as you were called to one hope when you were called; ⁵ one Lord, one faith, one baptism; ⁶ one God and Father of all, who is over all and through all and in all.

⁷ But to each one of us grace has been given as Christ apportioned it." (Ephesians 4:4-7)

Look at the declaration, *"But to each one of us grace has been given as Christ apportioned it." (Ephesians 4:7)* There is one Spirit, one Lord, and one God who is the Father of all, who is over all and through all and in all. And it is Jesus Christ who apportions Grace to each and every human being—to those who believe in Him and to those who do not believe in Him. *Ephesians 4:4-6* is talking about all mankind; and *Ephesians 4:7* is also talking about all mankind. And Acts 17:28 is talking about all mankind, as well; both the believers and the non-believers.

All human beings became living beings when God breathed His Grace into them; and there is no exception to this rule. That is why the Bible says in the following passage:

"The Spirit of God has made me; the breath of the Almighty gives me life." (Job 33:4)

Then, as we live and delve into situations in life and encounter challenges, we sometimes shine and increase in our Grace. And at other times, we stumble and lose Grace.

And just like in the popular saying that the cup is never half-filled; the life inside a human person is never half-filled. Whatever Grace is lost through a person's trials and tribulations, is made up with activated evil which promptly comes in and takes up its place. And while Grace fuels life and life's processes, activated evil brings decay to life; and impurities into life's processes.

And when we stumble but repent and make restitutions, Jesus Christ is right there on our side to increase our Grace and prop us up. Grace comes from the Father through His Son Jesus Christ. It is Grace that continues to hold up the human population throughout the world and throughout the ages. Without Grace, mankind would have been wiped out by the evil mankind has been committing.

There is no single human being who does not have the Spirit of God inside of him or her. The realities that we all know today to be our world and our universe are there because the Spirit of God remains in all things and keeps them what they are. Everything we have come to recognize with our senses remain what we have come to understand them to be because the Spirit of God remains in them and continues to keep them the same, for our sake.

Oftentimes, when we bash religion, we point to "reality" as our basis for doing so. What we call "reality" would not have been there in the first place if God had not chosen to put it in place—again, for our sanity. And this He did, not so we could use it to question His existence and the truthfulness of His word; but so that through the so-called "realities," we would come to know Him and obey Him.

Reality has many different layers and complexities. The things we easily observe in our environments are realities. Most of these are readily perceptible to just about everybody. Then there are other realities that take skills and time to unravel, say through science and logic. This group varies widely in complexity and requires certain expertise to fully comprehend.

And then, there are the mysteries of God which He graciously gave to us through the Bible, and through revelations. They are realities and require skills to comprehend. They require skills beyond what human science can teach anybody. The organic hardware that is the human brain simply does not have the capacity, by itself, to unravel this category of realities. The human person must rely on Grace, through faith in God, to decipher these mysteries.

Those who have trained themselves to comprehend the scientific and logical realities think of themselves as possessing the highest abilities to comprehend all realities, when it actuality, they are many degrees away from attaining the capability to fathom the more complex mysteries of God.

Because they lack faith in God, the window into God's higher mysteries is closed to their minds permanently. They would exhaust all that the organic fabric of their brains can do for them, and still not make any progress. And some will even develop mental illness and lose grip on common realities by trying to go beyond the threshold of their organic minds.

Only the Spirit of God could elevate any human being past that threshold. And to have the Spirit of God take you beyond the limits of the organic mind, you must have faith in God and remain in the truth of God. It is not something you do because you desire it. It is something that is given to you because the Spirit of God chooses to. The Grace that Jesus Christ apportions to all mankind is the only gateway into God's supernatural mysteries.

Many Christian teachers and preachers have walked away from the truth of God and ventured into the minefield of worldly knowledge in an attempt to blend the realities of science and logic with the word of God. And in so doing, they have adulterated the word of God, making caricature of the name of Jesus Christ; and stripping themselves of the power that comes from the Grace of God. They continue to rehash the wonders of yesteryears, when they could continue to operate in the power of the precious blood they profess day after day.

A true person of God does not feel threatened by the latest reality that is established by science. Human science is far too inferior to God's superscience, which He created and perfected, and employed in creating everything He had created and continues to operate. A true person of God stays with the word of God even when it seems contradicted by the human science.

God did not appoint you to spread His word because you understand everything. God appointed you to that position so God could teach you for the benefit of His people who will be listening to you.

It is a test from God to all those who profess faith in Him to be confronted with challenges from the world. Finding yourself in a position where your message faces severe challenges from science is a good thing! It is a good thing because your unshakable faith in the word of God causes the Holy Spirit of God who dwells within you to elevate you to that next level—beyond the capabilities of your organic brain. This is why Jesus Christ commanded all His disciples not to worry about what to say because the Spirit of God would put the words in their mouth anytime they must speak.

This is what the Bible calls, letting God be your teacher. And when God teaches you directly, not only does God make you understand what He is putting inside your mind through your spirit, He also helps you apply the information at the right circumstances. When God teaches you, your understanding of what He taught you is 100%. You would not forget any of it.

To substitute your so-called realities with the truth, so that you can, on your own, defend your faith, divorces you from the power of the Holy Spirit, and makes you vulnerable to failure and public ridicule. You are a servant of God and must only deliver the message you were given—not more, and not less! What you are eager to do for God is nothing that God could not do for Himself. So, stay with the message; uncompromised.

If the Christian leadership through the ages had remained adamantly faithful to God, there would not currently be thousands of versions of the Bible; or thousands of Christian denominations in the world. The Christian leadership over the ages had failed to understand that the Bible is the living word of God, and does not need human help to accomplish what God intended it to accomplish.

The earliest of these errors was the adoption of philosophy, and later science, as an alternate way of building and maintaining faith in God and His Christ. The church leaders, who instituted the practice and promoted it, and continues even up to now to celebrate the practice, do not realize that the practice would lead to the fulfillment of the dire prophecies in the Bible about the beast and the false prophet. That practice is the equivalence of Eve choosing to pursue the desires of her mind in favor of continuing to follow her spirit.

Eve chose her mind over her spirit. And the result was the activation of evil, and the decay which evil brought into the world. When the church decided in the middle ages that faith in God alone is not enough, the church unknowingly asserted that the mind is superior to the spirit. The world, through

Philosophy and Science, went the way of the mind, leaving the human spirit to languish.

By giving the mind the edge over the spirit, the church birthed and nurtured the second beast of Revelation, which eventually tore away from the church, and became poised on discrediting the church and demolishing the church's influence.

Religion can never become a substitute for godliness because religion has tainted itself by elevating the mind over the spirit! Every time the humans put the mind over the spirt, great evil got activated and unleashed into the world.

Eve did it and evil was activated in the world for the first time. She disregarded the commands of God, and chose to experiment as she was encouraged to by the devil. Adam, who was present at the temptation of Eve by the devil, did nothing to stop his wife, for whatever reason best known to him. And rot and decay set into everything in the natural world, including the human life and the human body, as a result of their disobedience.

The ancient Israelites catered more to their minds than their spirits—turning their backs on God, disregarding the commands and warnings of God, and instead, chasing after pleasure and material comfort—and they were ultimately driven away from their inheritance and scattered, in captivity, all over the world.

The church became fascinated with the education of the mind as an instrument for driving faith, and civilization. That practice led to improved cooperation among governments; improved goods and services; but anemic spirituality and the blending of the church and civil society. It has invariably led to the categorization of every western society as a Christian society.

However, in reality, many of these western societies, and the many other societies they have inspired around the world, have done everything in their powers, in the past so many years, to become societies of many religions and blended religious ideologies; making that unnatural marriage that has existed between the church and society a great liability to the church.

The line between godliness and excessive pursuits of modern luxuries has been totally blurred, making it virtually impossible for believers to reach a consensus on where to draw the line. It has also pitched the spiritual purist against secularist Christians who believe that civilization is a reward from God, and can never run counter to God. But it is exactly that kind of attitude that has sent mankind shuttling towards the prophecies of the End Time.

The same fascinating, but inferior, power of the mind is keeping many in the church from believing that the Last Day is really going to be the last day of everything as we see and know today. Instead, believers have resorted to the adoption of various different mixtures of spirituality and worldliness, in the guise of religion. May God have mercy on us!

In spite of all these, the Bible still remains the gateway to godliness and salvation, through the Grace which only Jesus Christ apportions to each and every human being—believers and non-believers alike. The Christian religious leadership has to make great adjustments so that its followers could give to Caesar what is Caesar's and to God what is God's. At this time, that line is blurred by greed, bickering about doctrinal correctness and the struggle for power, wealth and position, within the churches.

God is one! And the only way to have continued faith in God is by obeying God's every command as written in the Bible. And no one can obey something they do not believe in, in the first place. If you have issues with the way anything is written in the Bible, right there is your ticket to the disobedience of God. Any deviation from the truth of the Bible is unbelief and automatically positions one to be affected by activated evil.

That is why the Bible talks about sufficiently obeying God. Even those who profess their faith in God resort to picking and choosing what to believe and what not to. Anything you choose not to believe in the Bible positions your life to partake of the evil and its associated decay that are ever increasing in our world. It is automatic—unbelief equals decay! And you are the one bringing the decay into your own life, not God.

Everything that exists in whatever forms they exist do so through the Grace. And Grace is the Spirit of God. There is only one Spirit, and that Spirit gives life and all known and unknown characteristics to all of God's creations.

The Spirit of God is matter, time and space. The Spirit of God is gravity and all the forces that exist on the earth and in the entire universe. The Spirit of God is the constituent particles that make up everything that exists everywhere on earth and in the universe. The Spirit of God is the Grace of God which completely fills the earth and the universe and propagates through everything at the speed of God.

The Spirit of God is the glue that binds each of God's creation and the power that holds everything in its place. The Spirit of God is the force that initiates motion in anything, and the action and reaction that bears on every object in motion or under inertia. The Spirit of God is the energy that activates all things and directs them in the directions they must go. The Spirit of God is everything that exists in whatever forms they exist.

When the Bible says the following about the people in Noah's time, *"and they knew nothing about what would happen until the flood came and took them all away, that is how it will be at the coming of the Son of Man,"* *(Matthew 24:39)* humanity often thinks that people of that time were ignorant and had no way of predicting future events.

That is completely wrong. People of every age have sufficient human knowledge and wisdom that help them navigate through life and life's challenges. Otherwise, mankind would have become extinct. God has, through the ages, steered all humanity, and brought them to the present time. That humanity is oblivious of this fact, or in denial of it, changes nothing. God's truth is the only truth. And His truth will always stand.

God has had His hand in everything mankind has ever done in this world throughout mankind's history. A human being's basic existence is powered by God. A human being is not a living being until the Grace of God gets into him and powers him and helps him navigate through life. That is the only way life works. There is no other alternative available to any human being anywhere in this world or outside of it. Grace fills the universe and the earth completely, so wherever a human being finds himself, Grace is there to do what Grace does.

There are so many particles mankind has not been able to detect. God is limitless in His resourcefulness. No one can fathom the extent of His resourcefulness or accurately put it in words. Who God is, and the scope of His works, and His abilities, are simply beyond mankind's collective intelligence.

While Grace is the finest of all particles, spiritual or natural; other particles abound in both the natural and the spiritual realms. Every vocabulary that comes out of the mouth of God has its own distinctive form, its own characteristics, and its own functions; and brings forth into existence a new particle as different from every other particle as the night is different from the day.

Truth *(John 14:17)*; wisdom, understanding, knowledge, counsel & might *(Isaiah 11:2)*; leadership *(Numbers 27:18)*; supplication *(Zechariah 12:10)*; fire & judgment *(Isaiah 4:4)*; justice *(Isaiah 28:6)*; faith *(2 Corinthians 4:3)*; revelation *(Ephesians 1:17)*; righteousness *(Hebrews 12:23)*; glory *(1 Peter 4:14)*; and all other entities recorded in the Bible; are all distinct spiritual particles.

Everything that comes from God comes through Grace, thrives in Grace, and is controlled by Grace. In essence, it is Grace that gives life, form, and function, to everything in existence. Because Grace is God the Father, God the Son and God the Holy Spirit!

Love has a distinct form and function, just like light has a distinct form and function. And so does mercy, compassion, excitement, exhilaration, conscience, remorse, denial, fear, shame, anger, jealousy, envy, and all the other emotions that can be felt by human beings.

These are all spiritual particles. In their activated state, they evoke the right effect on natural beings. That is why the Bible talks about the spirit of dizziness *(Isaiah19:4)*; the spirit of despair *(Isaiah 61:3)*; the spirit of impurities *(Zechariah 13:2)*; the spirit of prostitution *(Hosea 4:2)*; the spirit of stupor *(Romans 11:8)*; the spirit of falsehood *(1 John 4:6)*; envy; fear; anger; jealousy; and so on and so forth. They are all discrete spirits, and infect the human persons as designed and commissioned by God.

While the human science, focuses exclusively on natural matters and energies which science is capable of detecting, quantifying and tapping into; a host of other particles exist that are beyond the scope of the human science, because they are spiritual in nature. This is why God said through the Prophet Isaiah:

""From now on I will tell you of new things,
  of hidden things unknown to you.
⁷ They are created now, and not long ago;
  you have not heard of them before today.
So you cannot say,
  'Yes, I knew of them.'" *(Isaiah 48:6-7)*

CHAPTER FOUR

Only You Can let the Devil into Your Life

If you leave your valuable on the side of the street without paying attention to it; or on a corner in a busy public space, and take your mind off of it for a while; when you go back later to get your valuable, chances are that someone has stolen it. Human beings, in their irrational state, steal from other human beings.

It happens all the time. Even to the point of people forcibly snatching things away from the rightful owners in broad daylight! And even the best government laws on theft and robbery does not discourage those who have conditioned themselves to steal from stealing. To them, it has become a force of habit. So, nobody can guarantee anyone that he is never going to get robbed.

God, on the other hand, guarantees that the devil cannot get to any human person unless the person personally lets the devil in. God restricted the devil from gaining access to the human spirit, because the human spirit has unobstructed access to the human mind.

Anything that gets into your spirit has unfettered access to your mind; and to the rest of your life. Therefore by keeping the devil away from your spirit, God guarantees that the devil cannot get into you, unless you consciously permit him.

Whatever gets into your spirit immediately begins to influence your mind. Only God has access to the human spirit, through Grace. So the only avenue available to the devil to get to a human being is through the human mind.

But unlike the human thief who can take from you without your permission; the devil cannot take from you without you first allowing him to. God designed your life in such a way that only your permission gives the devil access to your life.

And if you refuse to grant the devil entry into your life, the devil will depart from you, defeated. But if you, for any reason, grant access to the devil, the devil will gladly accept your welcome and take control of your mind. And once the devil comes into you, he seeks to increase his territory over your life, thoughts and actions.

The devil causes your body and your life to activate the evil particles which the devil brought to your life, thereby inundating your body and your life with rot and decay. Once the devil is in your life, he gets you to do things you ordinarily would not have allowed yourself to do. He works very fast, putting thoughts and ideas in your mind that would cause you to not only lose Grace, but to also increase the accumulation of activated evil in your life. His goal is to mess up your good standing with God, so the faster he acts on you the better his chances of getting you away from God, and away from your chance of salvation.

The higher the level of activated evil in your life, the higher the level of Grace required to restore your life to normalcy. Only repentance and remorse could get you the extra Grace you need to overcome your accumulated activated evil; and repair the damages in your body and your life caused by the activated evil.

If a man says to you: "Can I abuse your son?" Would you tell him to go ahead and abuse the child that you love? Would you not rather look at him in hatred and tell him to buzz of before you do something nasty to him? So, why do you readily give the devil the permission to do you bad? Why do you allow him to push you into things that would destroy your life?

This beast has no power over you, except the power that you grant him. Why then do you give him that power to mess up your life? He has absolutely no power, nor the authority, to force you into anything. Therefore, the same way you resist any human being that tries to get you into trouble is the way you must push back against the devil.

You must remember that a human being can inflict serious damage on you without your permission. Yet you successfully guard against people taking advantage of you and bringing damages on you. You securely tuck your money away so no one could get to it and steal it. You lock your valuables in safe places to avoid people taking them away from you.

Yet, you leave yourself wide open to the devil, even though your Father in heaven have graciously warned you against the deceit and vile ways of the devil. You allow yourself to enter filthy conversations with someone else's wife or husband knowing fully well that neither your spouse, or the person's spouse, would approve of what you two are doing. Yet you continue to engage in such a behavior because it brings you moments of ecstasy that fades away as fast as they come, which leaves you with a lifetime of regrets and a dark secret to carry around. Dark secrets and regrets rot your life away.

You suffer with envy over something that your friend has that you know very well that you cannot afford. You allow the devil to talk you into killing another human being or inflicting serious injuries on them so you can acquire that which you cannot afford. That emptiness in your life which you thought the thing you wickedly acquired will fill for you, will still linger around even after you obtained the stuff. Now you have more to worry about because activated evil is fast accumulating inside you and decimating your life.

In conclusion, God's love shuts the devil out of your life to start with. That is why the Bible says that you are wonderfully and fearfully made. Wonderfully made, because your anatomy, physiology and spirituality constitute a masterpiece of God's genius; and fearfully made, because no spiritual being could get to you, into you, or through you, without first seeking for your permission! And when, in your irrational state and foolishness, you fall for the devil's deception and allow the devil into your life; but repent, make restitution and turn to God, God's mercy restores you by buffering your body and your life with abundance of Grace.

When each human being is created by God, the person is created with abundance of Grace. That is why every human baby is holy. Each human baby has overflowing Grace to keep him or her from the devil and his minions. No baby is affected by activated evil because babies have enough Grace to keep out all activated evil from them.

And when we receive Jesus Christ as our Lord and Savior, God restores our Grace to the level it was when He first created us—an overflow of Grace. This overflow of Grace is tantamount to the Holy Spirit of God coming into you and dwelling inside you. Grace is not only replenished and reinforced by the Holy Spirit of God. Grace is parts and parcel of the Holy Spirit of God.

Grace is the purest and the most potent force on earth and in the universe. Grace operates everything. Grace holds everything in place as discrete things or beings. Grace holds everything together and Grace controls everything. Simply put: Grace runs the world and the universe and Grace is everything in the world and in the universe, because Grace is Jesus Christ who is one with God the Father.

Jesus Christ told us that being born again is being born of the Spirit. And He declared that we have to be like children to be born again. What Christ is talking about is real and practical, and not some make-belief that anyone can fake through. When you are born again, you cannot mistake that for something else.

Being born again brings a real peace into your life. You know that you know because it has been communicated to your spirit very clearly. You are now at peace with yourself and at peace with life, because your perception of Grace is perfected.

To be born again is to receive an overflow of Grace in your life. The overflowing Grace heals your tumultuous mind through the knowledge of God; and through God's revelations to you personally.

Contentment becomes an integral part of your life. Your trust in God rises exponentially. You learn to dismantle the numerous biases in your life which you picked up over the course of your life. The lives of all human beings begin to have great value in your mind, even the lives of those who still do you wrong.

And the words of the following passage begin to describe your everyday feelings:

"The law of the LORD is perfect,
  refreshing the soul.
The statutes of the LORD are trustworthy,
  making wise the simple.
[8] The precepts of the LORD are right,
  giving joy to the heart.
The commands of the LORD are radiant,
  giving light to the eyes.
[9] The fear of the LORD is pure,
  enduring forever.
The decrees of the LORD are firm,
  and all of them are righteous.

[10] They are more precious than gold,
  than much pure gold;

they are sweeter than honey,
  than honey from the honeycomb.
[11] By them your servant is warned;
  in keeping them there is great reward.
[12] But who can discern their own errors?
  Forgive my hidden faults.
[13] <u>Keep your servant also from willful sins;</u>
  <u>may they not rule over me.</u>
<u>Then I will be blameless,</u>
  <u>innocent of great transgression.</u>" (Psalm 19:7-13)

CHAPTER FIVE

Grace is Beyond Religion!

Grace is the science of everything that exists on the earth and in the universe, visible and invisible, natural and supernatural (physical and spiritual), living and non-living. Grace is everything! Everything God created is saturated with Grace; tied to one another by Grace; and tied to God by Grace. Grace is the essence of all beings; because Grace is God the Father, God the Son and God the Holy Spirit. Grace is the Trinity!

Grace is the substance in all of God's creation. Virtually everything that exists, exist due to Grace, and within Grace. Without Grace, nothing exists. Grace is the unlimited love of God. Grace is the boundless mercy of God. Grace is the infinite joy of salvation. Grace is the vitality of life. Grace is the illumination of the mind. Grace is the fuel that maintains vitality in all of God's creation.

The Bible makes it clear that God is talking to the entire world, believers and non-believers alike. God intends for every human being to understand the truth of his or her existence. And this is not a religious thing. It is the essence of our beings!

Here is one such verse in the Bible:

"Come near, you nations, and listen;
  pay attention, you peoples!
Let the earth hear, <u>and all that is in it</u>,
  the world, <u>and all that comes out of it</u>!" *(Isaiah 34:1)*

All of mankind belongs to God. Through His Grace, God takes care of the needs of every living thing. And to the humankind, God devotes special attention because all human beings have a special place in God's heart. God declared through Prophet Ezekiel that He takes no delight in the death of anyone, but delights in the redemption of all human beings:

"Say to them, 'As surely as I live, <u>declares the Sovereign Lord, I take no pleasure in the death of the wicked, but rather that they turn from their ways and live</u>. Turn! Turn from your evil ways! Why will you die, people of Israel?'" *(Ezekiel 33:11)*

"Rid yourselves of all the offenses you have committed, and get a new heart and a new spirit. Why will you die, people of Israel?" *(Ezekiel 18:31)*

Therefore, my writings are not religious writings. My writings seek to transcend religion, and reach all the children of God, who in their irrational state, have wandered away from God; thereby, walking away from life. God is trying to reach all mankind with His message. The same way God is trying to tear people aware from their maddening pursuits, God is also trying to tear people away from fanatical religion and human doctrines that lead no one anywhere.

God is trying to bring all humankind to the power of His Grace—the Grace that is abundantly in, and around, every human person; but has so far has been minimized by our respective and collective sins.

Grace does not only work in the context of religion. Grace works continually, inside and outside religion, because Grace is the very life that each and every human being has in them. Life does not stop for religion or for anything else in this world. Life is continually on because life is Grace and Grace is God the Father, God the Son and God the Holy Spirit.

The Bible says that when God opens His hand, the needs of all living things are satisfied. Therefore, it is not only those who have found Jesus Christ that currently live off the Grace of God: Everything that is alive in the world and the universe—spiritual and natural beings alike—are continually controlled and directed; monitored and corrected; replenished and restored; by the Grace of God. Here is that passage from the Bible:

*"<u>**All creatures**</u> look to you*
  to give them their food at the proper time.
[28] When you give it to them,
  they gather it up;
when you open your hand,
  they are satisfied with good things.

> ^{29} *When you hide your face,*
>   *they are terrified;*
> *when you take away their breath,*
>   *they die and return to the dust.*
> ^{30} <u>*When you send your Spirit,*</u>
>   <u>*they are created,*</u>
>   <u>*and you renew the face of the ground.*</u>*" (Psalm 104:27-30)*

Therefore, the Bible is beyond religion! And the Bible is also beyond human science, and all of humanity's specialized fields of study! The Bible is not a religious artifact. The Bible is the Book of life, dictated by God to give <u>all mankind</u> salvation and life.

The Bible contains the mysteries of God which God has posited for the enlightenment and salvation of all mankind. All human beings must embrace God to raise their threshold for Grace. Therefore, all human beings have to embrace the Bible because the Bible is the only Book God designed to show all human beings how to splurge into Grace.

To continue to present the Bible to the world as the Book of a religious faith is counter to God's intent. The Bible is the uncontested Book of God's truth. Every Christian, therefore, must take it upon himself or herself to present the Bible to the world for what the Bible really is—the Book of life, which is the Grace of God.

Allowing religious labels to prevent most of the world from abundance Grace is not the will of God.

Jesus Christ commanded His disciples to go to all people with the good news of the kingdom of God, baptizing everyone in the name of the Father, the Son and the Holy Spirit.

Religion is only a vehicle through which the good news of the kingdom could be taken to all the peoples of the world. Religion, in itself, is not the good news of the kingdom of God. That is why the good news got sidetracked in the Jewish religion; and Jesus Christ railed heavily on that religious establishment for walking away from the truth of God.

Christ's rebuke of the Pharisees and the leaders of the people was not a condemnation of the Jewish religion; but rather, a condemnation of their elevation of religion over the good news of the gospel.

It was not religion that got Moses to trust God and ask God to show him God's way so he—Moses—could always do what pleases God. Moses was not raised up in the Jewish religion. Moses was raised up as a true blood Egyptian, and cultured in the idol worshipping lifestyle of the Egyptian elites.

Moses grew up a Gentile king-in-training, and fully immersed himself in the business of exploiting the Jews. Up until his biological sister, Miriam, brought it to his attention that he was indeed Jewish, Moses did not waver in the exploitation of the Israelite slaves in Egypt.

It was not the Jewish religion that changed Moses's heart and turned him a defender of the Israelite slaves. It was Miriam's revelation that warmed up Moses's heart. The revelation—not the Jewish religion—compelled Moses to search his own heart for right and wrong. It was the conviction from Miriam's revelation that made Moses chose to interfere in the maltreatment of a Jewish slave by Moses's fellow Egyptian idol worshipper.

The revelation from Miriam transformed Moses into the tireless advocate of God that he became; and not the Jewish religion that brought that transformation about. That is why as the Israelites continued to struggle to hold on to their faith, Moses remained a rock, and continued to intercede on the behalf of the rest of them. His personal convictions—not his religion—has brought into display, the Grace that was abundantly in him, and all around him.

Cornelius, the centurion, did not have a religion, yet he continued to do what was right; touched by the sufferings and hardship of the people he was surrounded by. It was not the Jewish religion or any other religion that made Cornelius take up the burden of his neighbors. It was Cornelius's belief in doing what was right, and bringing peace to his own heart, that made Cornelius act. And he continued helping others, not hoping to receive anything in return. But his selfless acts made God sent Peter to receive Cornelius and his entire household into Jesus Christ.

It was not the religion of the Shunammite woman that endeared her to God's heart; and caused God to clear her path everywhere she turned. It was her trust in God, her contentment in life, her love of community, and her care and devotion to the man of God—all in the face of deprivation of a child—that endeared her to God's heart and brought her God's favor time and time again.

It was not religion that brought Ruth God's favor. It was Ruth's undying love for, and dedication to, her mother in-law who was old and tortured by life mishaps, that made Ruth a virtuous woman in the eyes of God and all mortals; and set her apart from all women throughout the generations. Ruth was a Moabite and was schooled in the religious experiences of her people. This fact actually caused some of the Israelite women to minimize Ruth when she returned to Israel with Naomi, her mother-in-law.

But out of faith, Ruth abandoned the way of life of her people, which she was so cultured in, and turned her back on her own parents, and followed old Naomi back to Israel, without any hopes for any inheritance. She maintained her sight on doing what was right, that nothing else mattered to her. She was determined to live and die with the older woman, out of a feeling of commitment and loyalty—the fate that brought them together, must have intended for them to remain together until death.

Therefore, it is not religion that God is focusing on, but the content of our hearts and our obedience to His commands. Labels are human fabrications because we humans could not see farther than our noses. So, we resort to practical things we can perceive and control. Unfortunately, God does not accept whatever we decide to do because we have all agreed to it. God looks at how well we conform to His commands, because His righteousness far exceeds our righteousness.

It was not religion that got the young Joseph through slavery in Egypt and kept him grounded throughout. It was Joseph's faith in the God of his fathers, Abraham, Isaac and Jacob, that got him trusting that God would come to his rescue—not only to free him from slavery, but also to make him into somebody. It was not religion that made Joseph shun his master's wife's sexual advances, but Joseph's devotion to doing that which was right. Joseph did not live in slavery in Egypt under the religion of his fathers.

He lived as a slave in the household of idol worshippers and could not afford to impose his religious beliefs over theirs. He humbled himself and respected their religious practices, while continuing to hold on to the faith his own ancestors has taught him. He even married the daughter of the chief priest of an idol worshipper and had two sons with her, who till today are counted as two parts of the inheritance of Israel.

And when God decided to bring Jacob and his descendants into the abundance of the land of Egypt and took Joseph from prison to the governor's mansion, God did not mandate that Joseph imposed his religion over those of his former slave masters. God simply overlooked the idol worship of the kingdom and established under Joseph a godly government that took care of all of its citizens and neighbors in a fair and equitable way.

God even allowed Joseph to continue to give allowances to the idol priests throughout the fourteen years, sparing the idol priests the sufferings and agony the rest of the population went through during the famine. God would have replaced the idol worshipping of that kingdom with, say, the Jewish religion since God had used the seven years of devastating famine to bring the kingdom to its knees. The people of Egypt were all thankful to Joseph and the wisdom he had received from his God that saved their lives during the famine.

Asking them to turn their backs on idol worshipping to survive the famine would have received a favorable response, if that were part of God's requirement for saving the people. But it clearly was not. And the people continued with their idol worshipping after the fourteen years of plenty and famine had passed.

This clearly demonstrates that it is not religion that God is looking for; but godliness! God cares about justice and equity more than God cares about religion. And Joseph's godly government in Egypt had both. That was why religion never mattered in that government.

The things God said to us in the Bible are not reserved to a select few who have already given their lives to Jesus Christ. Nor does being of the nation of Israel guarantee anyone's salvation. God is still very concerned with the welfare of the great majority of the humankind, who is still ignorant of His truth; and have been languishing in the scorching madness of life.

Anyone who gives their life to Jesus Christ has automatically signed on, to join Jesus Christ in getting His message of hope to the rest of mankind; and helping guide those that founder, back to Grace—which is life. Here is God calling all mankind away from sufferings and unrelenting wretchedness:

""Come, all you who are thirsty,
  come to the waters;
and you who have no money,
  come, buy and eat!
Come, buy wine and milk
  without money and without cost.
² Why spend money on what is not bread,
  and your labor on what does not satisfy?
Listen, listen to me, and eat what is good,
  and you will delight in the richest of fare.
³ Give ear and come to me;
  listen, that you may live.

I will make an everlasting covenant with you,
  my faithful love promised to David.
⁴ See, I have made him a witness to the peoples,
  a ruler and commander of the peoples.
⁵ Surely you will summon nations you know not,
  and nations you do not know will come running to you,
because of the LORD your God,
  the Holy One of Israel,
  for he has endowed you with splendor."

⁶ Seek the LORD while he may be found;
  call on him while he is near.
⁷ Let the wicked forsake their ways
  and the unrighteous their thoughts.
Let them turn to the LORD, and he will have mercy on them,
  and to our God, for he will freely pardon.

⁸ "For my thoughts are not your thoughts,
  neither are your ways my ways,"
declares the LORD.
⁹ "As the heavens are higher than the earth,
  so are my ways higher than your ways
  and my thoughts than your thoughts.
¹⁰ As the rain and the snow
  come down from heaven,
and do not return to it
  without watering the earth
and making it bud and flourish,
  so that it yields seed for the sower and bread for the eater,
¹¹ so is my word that goes out from my mouth:

It will not return to me empty,
but will accomplish what I desire
  and achieve the purpose for which I sent it.
[12] You will go out in joy
  and be led forth in peace;
the mountains and hills
  will burst into song before you,
and all the trees of the field
  will clap their hands.
[13] Instead of the thornbush will grow the juniper,
  and instead of briers the myrtle will grow.
This will be for the LORD's renown,
  for an everlasting sign,
  that will endure forever."" (Isaiah 55:1-13)

And this is not only a desire from God; it is assured, because the will of God always stands. Here are some passages about that:

"All the nations you have made will come and worship before you, Lord; they will bring glory to your name." (Psalm 86:9)

"Who will not fear you, Lord, and bring glory to your name? For you alone are holy. All nations will come and worship before you, for your righteous acts have been revealed." (Revelation 15:4)

CHAPTER SIX

Religion may hamper True Worship of God!

Jesus Christ is not just for the Christians in the world. Jesus Christ is for all of mankind; past, present and future! And all of humanity lives because He lives. And there is no exception to this reality, from the beginning of the world to the end of it. Jesus Christ said in the gospels, *"I and the Father are one." (John 10:30).* Therefore, whatever applies to God, the Father, also applies to the Son Jesus Christ.

The Bible says of Him,

"In the beginning was the Word, and the Word was with God, and the Word was God. ² He was with God in the beginning. ³ Through him all things were made; without him nothing was made that has been made. ⁴ In him was life, and that life was the light of all mankind. ⁵ The light shines in the darkness, and the darkness has not overcome it." (John 1:1-5)

"The Son is the image of the invisible God, the firstborn over all creation. ¹⁶ For in him all things were created: things in heaven and on earth, visible and

invisible, whether thrones or powers or rulers or authorities; all things have been created through him and for him. ⁱ⁷ He is before all things, and in him all things hold together. ¹⁸ And he is the head of the body, the church; he is the beginning and the firstborn from among the dead, so that in everything he might have the supremacy. ¹⁹ For God was pleased to have all his fullness dwell in him, ²⁰ and through him to reconcile to himself all things, whether things on earth or things in heaven, by making peace through his blood, shed on the cross." (Colossians 1:15-20)

And when the Bible says the following about God, the Bible is also saying it about Jesus Christ His Son, who is in the Father and the Father in Him, and who has been with the Father from the beginning: **'For in him we live and move and have our being.'** *(Acts 17:28)*

Therefore, every human being lives and moves in Jesus Christ. And in Jesus Christ, every single human being has his or her being, whether or not the person knows it and believes it. Your ignorance of that fact does not abolish its truthfulness. Look at this passage from the Bible:

"The wrath of God is being revealed from heaven against all the godlessness and wickedness of people, who suppress the truth by their wickedness, ¹⁹ <u>since what may be known about God is plain to them, because God has made it plain to them.</u> ²⁰ <u>For since the creation of the world God's invisible qualities—his eternal power and divine nature—have been clearly</u>

seen, being understood from what has been made, so that people are without excuse." *(Romans 1:18-20)*

Everybody that breathes and walks around on the surface of the earth exists and operates within Jesus Christ from whom all life originates; and in whom all lives are sustained. That a religion is ignorant of this truth does not change anything for the human beings who are a part of that religion.

Each and every one of the people that belong to all the different religions in the world, and those who do not belong to any religions, still have to answer to Jesus Christ as you can see from the following passage from the Bible:

"The God who made the world and everything in it is the Lord of heaven and earth and does not live in temples built by human hands. ^{25} And he is not served by human hands, as if he needed anything. Rather, he himself gives everyone life and breath and everything else. ^{26} From one man he made all the nations, that they should inhabit the whole earth; and he marked out their appointed times in history and the boundaries of their lands. ^{27} God did this so that they would seek him and perhaps reach out for him and find him, though he is not far from any one of us. ^{28} 'For in him we live and move and have our being.' As some of your own poets have said, 'We are his offspring.'

Therefore since we are God's offspring, we should not think that the divine being is like gold or silver or stone—an image made by human design and skill. In the past God overlooked such ignorance, but now he commands all people everywhere to repent. ³¹ For he has set a day when he will judge the world with justice by the man he has appointed. He has given proof of this to everyone by raising him from the dead." (Acts 17:30-31)

The ancient Israel symbolized the church in every way imaginable. That was God's grand design. God chose Israel over all the nations of the world so He could demonstrate His righteousness to mankind— the righteousness which is by faith only. And this was done by God to prepare the way for the coming of His Son Jesus Christ into the world; to make the atoning sacrifice for all mankind with His own life. And this is why the Bible says:

"But now apart from the law the righteousness of God has been made known, to which the Law and the Prophets testify. ²² This righteousness is given through faith in Jesus Christ to all who believe. There is no difference between Jew and Gentile, ²³ for all have sinned and fall short of the glory of God, ²⁴ and **all are justified freely by his grace through the redemption that came by Christ Jesus.** *²⁵ God presented Christ as a sacrifice of atonement, through the shedding of his blood—to be received by faith. He did this to demonstrate his righteousness, because in his forbearance he had left the sins committed beforehand unpunished— ²⁶ he did it to demonstrate his righteousness at the present time, so as to be just*

and the one who justifies those who have faith in Jesus." *(Romans 3:21-26)*

What Israel was in the Old Testament is what the church is in the New Testament. Unfortunately, in the middle ages, the church did not understand it that way. And that misunderstanding made the church try to wipe the Jewish religion off the map.

The church was not intended by God to be large religious organizations with their own unique set of doctrines which meet in huge buildings to listen to sermons, sing hymns and made exclusive claims about being the only approved church on the earth.

It was collective human wills that turned the church into what we recognize today as the church— hugely fragmented and bickering over whose doctrines are the true doctrines approved by Jesus Christ.

Jesus Christ prayed to His Father before He gave His life in atoning sacrifice for mankind. He asked the Father to keep the church as one so that the church will be united in all it does. Jesus Christ made sure that His apostles got His message. After He rose from the dead, He came to them and while they were eating, He dramatically handed the leadership of the church to Apostle Peter, commanding Him to take care of His sheep and to <u>feed</u> His sheep.

Jesus Christ did not only charge Apostle Peter with leading the church, and helping the church to maintain its spiritual fervor. He also commanded Apostle Peter to feed the church. That is the kingdom economy right there.

Jesus Christ did not only intend His church to meet and spread the gospel. He intended for the whole church to become a viable economy, with self-sustenance. The following was true of the ancient Israelites when in their disobedience of God they were attacked and scattered all over the world in the Gentile lands:

*"Then Haman said to King Xerxes, "**There is a certain people dispersed among the peoples in all the provinces of your kingdom who keep themselves separate**. Their customs are different from those of all other people ..." (Esther 3:8)*

Jesus Christ had intended for all who believe in Him to live as one people, and in one spirit, throughout the face of the earth. The kingdom of God was intended to be the dominating concern of all who believe in Jesus Christ, regardless of their political, geographical, cultural or ethnic background.

Although they are dispersed among the peoples throughout the world, they are supposed to keep themselves separate through their common faith in Jesus Christ, and to maintain the hope in the salvation that they seek in Jesus Christ—the salvation they must live for and be prepared to die for.

Intermarriage with the surrounding nations was banned for the Israelites:

"Therefore, do not give your daughters in marriage to their sons or take their daughters for your sons. Do not seek a treaty of friendship with them at any time, that you may be strong and eat the good things of the land and leave it to your children as an everlasting inheritance.'" (Ezra 9:12)

Interrelationship with nonbelievers was also banned for the church in all the communities the church exists:

"Do not be yoked together with unbelievers. For what do righteousness and wickedness have in common? Or what fellowship can light have with darkness?" (II Corinthians 6:14)

And this is without exception.

All believers of Jesus Christ everywhere in the world should enjoy a common bond as one family through the Spirit of love that brings them together and holds them together. Believers across different geographies and different political landscapes are expected to feel closer to one another than they feel to their relatives who shun faith in Christ and ridicule the things of God. That is why Jesus Christ declared in the gospels:

"'Do not suppose that I have come to bring peace to the earth. <u>I did not come to bring peace, but a sword.</u> ³⁵ For I have come to turn

"'a man against his father,
 a daughter against her mother,
a daughter-in-law against her mother-in-law—
³⁶ <u>a man's enemies will be the members of his own household.</u>'

³⁷ **"<u>Anyone who loves their father or mother more than me is not worthy of me</u>; anyone who loves their son or daughter more than me is not worthy of me. ³⁸ <u>Whoever does not take up their cross and follow me is not worthy of me.</u> ³⁹ Whoever finds their life will lose it, and whoever loses their life for my sake will find it.**

⁴⁰ **"<u>Anyone who welcomes you welcomes me, and anyone who welcomes me welcomes the one who sent me</u>.** *⁴¹ Whoever welcomes a prophet as a prophet will receive a prophet's reward, and whoever welcomes a righteous person as a righteous person will receive a righteous person's reward. ⁴² And if anyone gives even a cup of cold water to one of these little ones who is my disciple, truly I tell you, that person will certainly not lose their reward.""* (Matthew 10:24-42)

Natural biological families may differ over whether or not to accept Jesus Christ as their Lord and Savior. And those who hold out will see those who accept the faith as traitors. They will do everything in their powers, fighting the believers, in an attempt to set their minds right again. Brothers and sister would disagree over Christ. And so will other relationships within a family.

But people from different societies and nations, who join their hearts with the Lord Jesus Christ, would come together as one family with mutual love and compassion. These are the people that make up the church. They should live together as brothers and sisters in the Lord, feeling for one another and praying for each other's strength in keeping the faith and surviving attacks from the nonbelieving forces. They should support one another without prejudice.

But through bickering and struggling for personal power and recognition, the church is in disintegration; broken up into a million different pieces, each with prideful leadership and its own greed.

God provides for His people everything they need to sustain themselves. God personally builds His own economy. He does not rely on what the world has built. He allows His people to borrow from the world only that which God does not intend to provide directly to His people, like when He asked the Israelites to ask for, and receive gold, silver and clothing from the Egyptians before they set out on their journey to the Promised Land.

In the wilderness, when Moses was leading the Israelites to the Promised Land; and God decided that is was time for the Israelites to start forging their own spiritual, economic and political life; God instructed Moses on how to go about doing everything: from organizing the leadership of the people to developing the skills required for the various trades. Here is the passage from the Bible:

"Make sacred garments for your brother Aaron to give him dignity and honor. ³ <u>Tell all the skilled workers to whom I have given wisdom</u> in such matters that they are to make garments for Aaron, for his consecration, so he may serve me as priest." (Exodus 28:2-3)

"Then the LORD said to Moses, ² "<u>See, I have chosen Bezalel son of Uri, the son of Hur, of the tribe of Judah, ³ and I have filled him with the Spirit of God, with wisdom, with understanding, with knowledge and with all kinds of skills</u>— ⁴ to make artistic designs for work in gold, silver and bronze, ⁵ to cut and set stones, to work in wood, and to engage in all kinds of crafts. ⁶ <u>Moreover, I have appointed Oholiab son of Ahisamak, of the tribe of Dan, to help him. Also I have given ability to all the skilled workers to make everything I have commanded you:</u> ⁷ the tent of meeting, the ark of the covenant law with the atonement cover on it, and all the other furnishings of the tent— ⁸ the table and its articles, the pure gold lampstand and all its accessories, the altar of incense, ⁹ the altar of burnt offering and all its utensils, the basin with its stand— ¹⁰ and also the woven garments, both the sacred garments for Aaron the priest and the garments for his sons when they serve as priests, ¹¹ and the anointing oil and fragrant incense for the Holy Place. They are to make them just

as I commanded you."" *(Exodus 31:1-11)*

"The LORD said to Moses, ² **"Tell the Israelites to bring me an offering. You are to receive the offering for me from everyone whose heart prompts them to give.** *³ These are the offerings you are to receive from them: gold, silver and bronze; ⁴ blue, purple and scarlet yarn and fine linen; goat hair; ⁵ ram skins dyed red and another type of durable leather; acacia wood; ⁶ olive oil for the light; spices for the anointing oil and for the fragrant incense; ⁷ and onyx stones and other gems to be mounted on the ephod and breastpiece.*

⁸ "Then have them make a sanctuary for me, and I will dwell among them. ⁹ Make this tabernacle and all its furnishings exactly like the pattern I will show you." *(Exodus 25:1-9)*

God provides for His people everything they need to be successful in the things He had commanded them which they are obedient to. And if they lack anything, God supplies it to them directly. God does not make His people rely on the world to build up the things that are necessary in His kingdom. God had a set and matured economy for His kingdom. There is currently lack in the church because of selfishness, incompetence, gross mismanagement and outright abuse. This is why God said to His people:

"However, there need be no poor people among you, for in the land the LORD your God is giving you to possess as your inheritance, he will richly bless you, [5] if only you fully obey the LORD your God and are careful to follow all these commands I am giving you today. [6] For the LORD your God will bless you as he has promised, and you will lend to many nations but will borrow from none. You will rule over many nations but none will rule over you.

[7] If anyone is poor among your fellow Israelites in any of the towns of the land the LORD your God is giving you, do not be hardhearted or tightfisted toward them. [8] Rather, be openhanded and freely lend them whatever they need. [9] Be careful not to harbor this wicked thought: "The seventh year, the year for canceling debts, is near," so that you do not show ill will toward the needy among your fellow Israelites and give them nothing. They may then appeal to the LORD against you, and you will be found guilty of sin. [10] Give generously to them and do so without a grudging heart; then because of this the LORD your God will bless you in all your work and in everything you put your hand to. [11] There will always be poor people in the land. Therefore I command you to be openhanded toward your fellow Israelites who are poor and needy in your land." (Deuteronomy 15:4-11)

Sometimes, God waits for the people to ask of Him so He could demonstrate His faithfulness to them. But when the people forget God, and instead seek from the world, God never fails to show His dissatisfaction with the people.

After Israel had been established as a viable

spiritual, economic and political nation, and had so much material abundance, the Israelites got into themselves and started to blend in with the world around them, even copying some of the nations and seeking their expert advice in various areas of life. God felt humiliated at this and reacted with anger towards Israel. Through the Prophet Jeremiah, God communicated His anger to the people of Israel:

"Then the word of the LORD came to me. ⁶ He said, "Can I not do with you, Israel, as this potter does?" declares the LORD. "Like clay in the hand of the potter, so are you in my hand, Israel. ⁷ If at any time I announce that a nation or kingdom is to be uprooted, torn down and destroyed, ⁸ and if that nation I warned repents of its evil, then I will relent and not inflict on it the disaster I had planned. ⁹ And if at another time I announce that a nation or kingdom is to be built up and planted, ¹⁰ and if it does evil in my sight and does not obey me, then I will reconsider the good I had intended to do for it.

¹¹ "Now therefore say to the people of Judah and those living in Jerusalem, 'This is what the LORD says: Look! I am preparing a disaster for you and devising a plan against you. So turn from your evil ways, each one of you, and reform your ways and your actions.' ¹² But they will reply, 'It's no use. We will continue with our own plans; we will all follow the stubbornness of our evil hearts.'"

13 Therefore this is what the LORD says:

"Inquire among the nations:
 Who has ever heard anything like this?
A most horrible thing has been done
 by Virgin Israel.
14 Does the snow of Lebanon
 ever vanish from its rocky slopes?
Do its cool waters from distant sources
 ever stop flowing?
15 Yet my people have forgotten me;
 they burn incense to worthless idols,
which made them stumble in their ways,
 in the ancient paths.
They made them walk in byways,
 on roads not built up.
16 Their land will be an object of horror
 and of lasting scorn;
all who pass by will be appalled
 and will shake their heads.
17 Like a wind from the east,
 I will scatter them before their enemies;
I will show them my back and not my face
 in the day of their disaster.""(Jeremiah 18:5-17)

It was this conviction that led the leaders of the Catholic Church in the middle ages—when it was the only church on the earth—to valiantly resist secular influences over the church, but rather hold on to the truth that the church was designed by God to have temporal powers over the believers of Jesus Christ, and not just powers over their spiritual life alone. The church essentially ruled over the world and brought civilization to the whole world through its leadership.

The church's undoing, however, was that the church extended its temporal powers over all the people of the world; and not just over the believers of Jesus Christ—partly because the surge of the secular governments pushed the church into new and uncharted territories.

The church simply did not have the wherewithal to transition from managing the affairs of the world to simply managing the affairs of the believers of Jesus Christ the same way it had successfully done with the whole world before it was minimized by emerging political powers in Europe.

Once the secular powers in Europe rose over the church and constrained the church within a much smaller geography than the church was used to; and a string of corrupt Popes took the reign of the church leadership; the church essentially gave up temporal powers over everybody, and concentrated exclusively on the spiritual health of the believers. That was a big loss for the faithful; a misdirection for the church; and a great setback for the kingdom of God on the earth.

The kingdom economy just vanished from the face of the earth. The church had been essentially reduced to just the equivalence of the Temple in Jerusalem, when Jesus Christ had intended the church to be an upgraded version of the ancient

Israel, with a viable economy where poverty could not take any footing, because through the kingdom economy, God would amply supply to His people.

Ancient Israel was a real society. The church also is designed to be a real society. America is proudly touted as a melting pot because of how people from different countries, cultures and languages converge to the land and became a people with not only a common geographical boundary and common political ideology; but also a common spirit as one nation and a common need for fairness and justice.

If a political system—through the will of men—could successfully provide for the needs of its people, why is it so difficult for the church of God to operate in a similar fashion; especially since the first century church demonstrated that it is practical? The first century church was oppressed more than the church is oppressed in our current world; yet they lived according to the model Jesus Christ intended for His church on the earth.

Apostle Peter and all the others, most of whom had families, focused on the kingdom of God; believing what Jesus Christ commanded:

"But seek first his kingdom and his righteousness, and all these things will be given to you as well." *(Matthew 6:33)*

They obliged and their families were taken care of as well. There is no single reference in the Bible about any of the apostles accumulating big financial wealth and leaving for their children big monetary inheritance. Yet Christian leaders today believe in the pursuit and accumulation of personal wealth for their own use just like everybody else routinely does.

In the time of the apostles, those who had things of value were selling these things and bringing the money into the church to help provide for the swelling church membership—men and women and their children who left their homes and livelihood to embrace the word of God and turn to godly living: Many lost their properties to the local authorities. Many were driven out of their homes and banished altogether. Many were murdered or maimed because they refused to denounce the faith.

This is the type of faith Jesus Christ wants to see in His believers—the people who worship God in truth and in the Spirit. These pioneers placed the kingdom of God ahead of their own lives, their security and their personal comfort.

Nobody who worships God and proclaims His name only when it is convenient to do so worships God in truth and in the Spirit! Neither does a person who only recounts some learned praise or worship of God. See this passage from the Bible:

"The Lord says: These people come near to me with their mouth and honor me with their lips, but their hearts are far from me. Their worship of me is based on merely human rules they have been taught.'"
(Isaiah 29:13)

God condemns that kind of worship. The worship of God must come from the heart. Even if it involves prayers or praises one has been taught, its recitation has to come from the heart, or it becomes mere literature, intended to avoid hell than a demonstration of love for Jesus Christ and God the Father. That is why Jesus Christ said that those who love their lives on the earth will lose it; and that those who hate their lives will have eternal life.

It was in the face of people bringing in their life's treasures into the church to help support the gospel of God that Ananias and Sapphira decided to sell their own property as well, and give the funds to the church. But sadly for them, they forgot they were dealing with the Spirit of God, and not just human leadership of the church. They decided to lie to Apostle Peter about their own money—which they did not have to give if they chose not to—and they lost their lives in the process.

Their experience helps to establish that what the apostles and the early disciples sacrificed was the will of God. God had personally signed off on what they were doing—selling what rightly belonged to them and giving the money to the church to help support their brothers and sisters in the church.

And the church made sure that the Spirit of God oversaw the distribution of these blessed gifts among all the believers to ensure there was no poverty within the church. The apostles directed the believers to appoint from their midst men who were honest and filled with the Spirit of God to oversee the distribution of the gifts among the people. The church obeyed and the result was phenomenal because everything they did was approved by God.

If God did it for them, God stands ready to do it for the church today. But the church these days is more interested in protecting its wealth than in getting God to work for the believers through the church's resources.

Currently the church is so fragmented and each fragment is more into maintaining its viability in an earthly sense, than it is in taking a plunge of faith and trusting in the God whose name it uses in the first place.

Everyone is interested in having the assurance that he could take care of himself at any given time, even when the society as a whole is faced with hard times. And this is largely because the church had misrepresented the gospel of Jesus Christ.

The church (and religion) was designed by God to be for the people what the light house is for the sailors. The church is supposed to speak to everyone that casts an eye on it; anyone whose heart desires God and remediation from the problems of life.

The church was designed to assure the hurting that the church is a place to come to and receive help for their sufferings and life's troubles. The church building is supposed to represent real hope for the people whose hearts are troubled and whose lives are suffocating them to death.

The church was designed to always have believers who trust the Lord and are always prepared to assist anyone who walks through its doors.

The church was designed to serve the troubled and the outright disenfranchised. And God stands ready to help any church that takes this approach. God is never short of resources—all resources in the whole wide world belongs to God and He makes it available to everyone who follows His commands in dealing with the business of the people.

That churches are in lack today and have insufficient resources to help their members and the communities they serve is because the churches have broken faith with their God. They frequently subscribed to the systems of the world in solving problems within the church—financial, judicial, economic, and even spiritual.

The churches ascribe to the economic principles of Wall Street in preference to the economic principles of the kingdom of God, generously spread out all over the bible.

The churches place more trust in the promises of cutting edge medical systems than they trust in the healing powers of God which God promises to those who truly believe. And when modern medicine fails them, they finally come to God, looking for a miracle healing.

To those who truly believe in God; and hold onto every word and promise of God with all their lives, God works miracles in their lives every day.

The church was designed to experience daily miracles; and not one miracle in every blue moon. That the miracles are barely happening today is a reflection of how much the church loves God and obeys His commands.

The church has drifted so far away from God, and has become contended with being at par with the rest of the other world's institutions and religions.

The church is greatly missing out on God's wonders, and God is beckoning for the church to return to Him and experience His compassion and great love; just like He did with ancient Israel.

And when the church's fortunes are diminished and the church could no longer pay its bills, it loses its remaining assets altogether. Many church buildings in Europe and American have been converted into bars, night clubs and apartment buildings.

This is making God look like He could not provide for His people when in actuality it is the people and their greed and divisiveness that are leading the church down the tube.

Unfortunately, however, it is not only the congregation that goes under that looks bad; all of Christianity looks bad and loses credibility.

These days, prosperity Christianity has got everyone focusing on generating enough personal wealth for their families and their retirement. Greed and occupying positions of importance within the society keep believers pursuing earthly wealth as much as everyone else.

The church celebrates the earthly accomplishments of church members in ways that encourage others within the church to set their eyes on such earthly gains to feel blessed; when the church should be celebrating the benefits those earthly achievements contribute in alleviating the life's problems of the people within the church. Those are the only meaningful benefits of individual earthly accomplishments to the kingdom of God.

God had intended religion to foster godliness among the human beings, but He never intended religion to become more important that godly living among believers.

It is mankind that raised religion to the level it took over from the true virtues God intended for mankind's return to holiness. Every preacher out there is denouncing religiosity in words, yet every single one of them is practicing the religion that overshadows godliness. They are still all quarrelling about words.

It is not what we say that makes us different from the others. It is what we truly practice, and what we ultimately get out of our efforts, that set us apart from others.

Any Christian leader who preaches his denomination's doctrines as being superior to all the others out there, and uses that to measure the success or failure of his parishioners, is definitely more interested in religion than he is in godliness. He can bash religion all he wants, but he too, favors religiosity over the truth of God.

Religion is only important to God if it compels the church members to pursue godly living to the point their lives become light and salt to their neighbors. Anything outside of this is simply religiosity and leads believers away from the truth of God. It makes believers go through the motion, even in their charity, unaffected by the truth of the gospel.

Preachers like this do not and cannot show real love to their neighbors because everything they do is directed by their minds; and not by their spirits.

The mind is not capable of producing love because love dismantles all logic. Love is spirit-driven, and reacts instantaneously to suffering and deprivation.

Religiosity cannot promote love, because religiosity requires that the church doctrines be taken into consideration before a believer reacts to situations. Everything is measured and checked against the church teachings before a decision is reached on whether to intervene or not.

God declared through His Son Jesus Christ that those who worship God must worship Him in truth and in the Spirit. How then did we allow religion to overshadow the true worship of God?

And the best worship to God is obeying His commandments. Every single time you consciously do what God says to do; or the occasion arises and you, out of obedience to God, abstain from what God says to abstain from; you are worshipping God—showing your love and reverence for Him.

We are presented with the opportunities to worship God much more in our private lives that we are in the short time we spend in church services on

Sundays. That is the way true worship happens because our true worship of God provides real values to other human beings that have the chance to experience us.

Our worship of God should help restore the people around us. This is the true worship of God. And that is what Pope Francis recently recommended to the Roman Catholic priests and parishioners across the world.

God does not want our worship of Him to be limited to the church buildings. God wants all believers to worship Him every minute of their lives; everywhere they find themselves; and in all situations they get into. Jesus Christ said to the Samaritan woman at the well:

"Woman," Jesus replied, "believe me, a time is coming when you will worship the Father neither on this mountain nor in Jerusalem. ²² You Samaritans worship what you do not know; we worship what we do know, for salvation is from the Jews. ²³ **Yet a time is coming <u>and has now come</u> when the true worshipers will worship the Father in the Spirit and in truth, for they are the kind of worshipers the Father seeks.** *²⁴ God is spirit, and his worshipers must worship in the Spirit and in truth."*
(John 4:21-24)

Apostle Paul declared:

""The God who made the world and everything in it is the Lord of heaven and earth and does not live in temples built by human hands. ^{25} And he is not served by human hands, as if he needed anything. Rather, he himself gives everyone life and breath and everything else. ^{26} From one man he made all the nations, that they should inhabit the whole earth; and he marked out their appointed times in history and the boundaries of their lands. ^{27} God did this so that they would seek him and perhaps reach out for him and find him, though he is not far from any one of us. ^{28} 'For in him we live and move and have our being.' As some of your own poets have said, 'We are his offspring.'

^{29} "Therefore since we are God's offspring, we should not think that the divine being is like gold or silver or stone—an image made by human design and skill. ^{30} In the past God overlooked such ignorance, but now he commands all people everywhere to repent. ^{31} For he has set a day when he will judge the world with justice by the man he has appointed. He has given proof of this to everyone by raising him from the dead."" (Acts 17:24-31)

Christianity has become exactly like the Judaism of two thousand years ago. Believers have mostly substituted godliness with religiosity. And such an adulteration of the faith was sharply rebuked by Jesus Christ everywhere He went while He was in the world. Jesus Christ openly rebuked the religious establishment and its leadership for holding God's people hostage, and leading them into their destruction and hell; instead of their salvation and eternal life.

But before anybody thinks that God is condemning religion, we should all be reminded that God started religion and continues to encourage believers to stick with religion. God gave to Moses a set of requirements for the people of Israel so they could work with one another and promote the affairs of the people over their personal ambitions. That is what religion is all about.

God's problem with religion is that religious people often abandon the true premise of religion and replace the pursuit of godliness with religiosity. And this problem always starts with the religious leaderships who place their needs above the need of the people; and profit above the need of serving God's purpose in the lives of God's people.

Jesus Christ lashed out at them from every angle to get them to change their practices and subscribe entirely to the word of God in the Scriptures. Yet Jesus did not tell believers to get away from these religious leaders and the Temple life in order to set their lives right before God.

Instead, He commanded them to listen to these leaders who were entrusted with the safeguard and dissemination of God's eternal word. However, Jesus Christ warns believers to not only be skeptical of religious teachings that contradict the

Bible or deviates even very slightly from it, but to reject such false teachings completely.

Christianity has drifted from the way it was designed to be by Jesus Christ; and has become in every way imaginable like the Judaism of the first century. These days, Christian leaders are more interested in protecting their income and their elevated positions than they are in advancing godliness in the lives of the believers; who dedicatedly invest time and money into the church in the first place to allow these church leaders to carry out the affairs of the church.

God established religion for the betterment of the lives of the people of God, and not so that a religious name is established in neon lights, and the church leaders famed. That is religiosity and serves only the interest of the people who promote it; and not the lives of the believers whose support established the church in the first place. Religiosity is a disservice to the people of God who throw their meager earnings to prop up the church; and a rebellion against God.

Adam and Eve were living holy lives in the Garden of Eden, which was the holiest place on the earth, when the devil walked in and corrupted their hearts, causing them to rebel against God. So, the devil is not afraid to come into a community of God's people and lure its leadership into doing a disservice to the people and disobedience to God.

When religious leaders turn from leading worships, to become the objects of worship themselves, they are stealing glory away from God—to their own detriment.

Jesus Christ commanded the people of God to listen to every word the teachers of the law spoke to them; but not to copy the way these leaders apply the law of God to their own lives. Jesus Christ reminded the people of God that, in spite of the poor performances of these leaders, the religious leaders were charged with teaching the people the word of God. Why? Because these religious leaders, who place religiosity ahead of the job they were charged with, were operating from the seat of Moses—the humble servant of the all-powerful God through which God gave the law to His beloved Israelites.

Jesus Christ was sure to sanction religion when He personally, after His resurrection from the dead, charged Apostle Peter to feed His sheep and take care of His sheep. His sheep is the church, an intricately woven association of those who love Him and believe in Him—the believers He commanded to seek first the kingdom of God so that everything else will be added unto them.

These believers obliged and Apostle Peter also obliged, and the faith was set on fire. And the result is the church that has grown, and today is everywhere there are people on the face of the earth.

Religion is definitely relevant to godliness, but religion must never prostitute itself as a replacement for godliness. Religion is a marker intended by God to attract the poor and the disenfranchised; the hurting and the suffering; and all those who need a renewed spirit and a renewed heart that is full of life.

Religion is designed to bring these people together and provide them with a place and the spiritual and material resources they need to get their lives together in Christ Jesus; and start their journey to godliness and holy lives.

Religion is a place that is designated to support and carry the people of God into the lives God designed for them. Religion is supposed to be a place filled with love; a place that embraces all who enter through its door with open arms, immeasurable love, great compassionate care and abundant mercy.

And religious leaders are servants of God placed in the church by God to provide restorative services to the people of God and serve their every need; to help them in their journey to godliness and holy living.

Religious leaders are not intended by God to become entrepreneurs of big religious institutions who deploy Wall Street ideologies in maximizing the profitability of their religious ventures. All religious ventures must have the people of God as their collective beneficiaries.

Otherwise, they become self-serving worldly ventures that disguise themselves as religious outfits. Such is idolatry and misuses the name of God and of His Son Jesus Christ.

If the first century followers of Jesus Christ placed the love and the devotion of their fellow believers ahead of their own personal securities and sold their personal properties and gave all the proceeds to the church for the support and maintenance of all the people of God; how much then do we owe it to God's people now, to use most of the proceeds from our endeavors as church leaders to support the people that support the church?

God's economy is a real economy, and far superior to any world economy. But the reality in our world today is that the church economy is piggy-banked onto the world economy, and could not stand on its own because it has no foundation to stand on.

Most of the money that comes into the church is today spent on maintaining the high-flying religious ministries, and returns little or nothing to the people that come together week after week to support the religious endeavors.

God promises so much to the people that believe in Him and obey His commands. Much of this is to come from the church economy. Instead, nothing comes from the church economy because it does not even exist. As I said in the beginning, the church is to the New Testament what ancient Israel is to the Old Testament. God encouraged Israel to develop a real economy, and have its people serve within that thriving economy.

That is what sustained Israel and helped the society blossom into a viable economy. And that was supposed to be the way of the church. Israel supported and sustained its people. The church is supposed to support and sustain its people.

Israel was not prosperous out of the charity of its people but rather flourished under God's economic principles as written in the Law of Moses. The church is not expected to proper out of the charity of the believers of Jesus Christ, much less the charity of non-believers.

Rather, the church is supposed to flourish under the kingdom economic principles as taught by Jesus Christ and exemplified by His apostles and His disciples of the first century Christianity. God's kingdom economy is a real economy and is powerful enough to prosper all of God's people on the earth, if we believe and fully obey God.

But when we do not trust God enough to obey His commands as He commanded them, and set our eyes on the prize; but instead put our hopes into world's economic principles and hustle with the rest of the world according to the world; then the church could only thrive on charity, and the people of God would live in impoverished state, with everyone looking after his own worldly family, and barely meeting its needs.

The Bible declared that any hope that is seen is not hope at all. Our faith should not be based on what the world teaches but on the promises of God that are strewn all over the Bible. God is not a man that He would fail to honor any of His promises to His people. But as is always the case with everything God, only faith can unlock the power of His mysteries for us and inundate our lives with the abundance Jesus Christ promised to all who believe in Him.

When we do not believe that only our faith in God can heal us, but instead places more trust on human medicine and our ability to pay for the medical services; then our healing would not come from our faith in God but from human medical systems.

When our hope for security comes from the money and properties we have stashed away, and not from our confidence in God and all He had promised us; then God's protection would not be our first line of defense but rather, our earthly wealth would lead the way as best it could. In the end, it always falls short of the target.

Similarly, when the church operates on the world's economic principles, the return on its investments is directly proportional to all the other earthly returns on investments—hit or miss! But when the church of God operates on the kingdom's economic principles, the return on its investment is the financial windfall God promises in the Bible—the church could not contain the favor God will pour on the church. And as such the church will have all the resources it needs to support the needs of all the people in the church.

Like Israel, the church is to help its people tap into the kingdom economy by eliciting the expertise and professional services of believers who are versed in the kingdom economic principles. Every church of God is supposed to be a thriving economy for those who worship in it. There should be no poverty in the church, as God declared to the Israelites, because God's favor will abundantly supply all the believers in the church.

Every preacher aspires to get on the Television with his gospel in order to gain in popularity and recognition. Everybody copies everybody else and rushes to get it on the air or in print before the next guy does; sometimes even before the person that receives the message from God can get his message out. Because of this attitude, preachers copy even the wrong message in the attempt to be the first.

We rewrite and re-interpret the Bible to match the facts human science and the world believes, in an attempt to make the gospel of Jesus Christ appeal to the world. The message of the cross is fast becoming our individual messages, tailored to make the biggest draw for us than to actually lead people to godly living.

Preachers attract and fool around with other people's wives; instead of discouraging their followers from engaging in atrocities like that. Many steal from the church and gamble with the lives of the people they are supposed to mentor and guide spiritually and otherwise. Anybody who can speak fluently becomes an overnight sensation in the delivery of TV evangelism, just to fill his pocket with contributions coming from believers. Preachers are fast becoming television salesmen who are experts in selling lies to the people—those who clamor around them—because they are telling them what they want to hear.

If there is this much iniquities in the church, does the word of God that we represent and preach actually have any impact whatsoever on our lives? Or has the spreading of the gospel become mere vocations and sources of financial rewards to all of us?

Evangelists now measure the success of their evangelism by the number of people that show up at their crusades. They count all of those who respond to their alter calls as those they led to God; not considering that many among these people might have been a part of the Christian faith for a long time, but answered the alter calls to show their allegiance to God.

The only count of conversions that matters to God is the count God makes Himself. God's count is the true measure of the success of anyone's discipleship. God does not tell any of us what these numbers are, but He implicitly fills us with the satisfaction that we are making the right progress in His kingdom.

That is why Apostle Paul reminded us in the Bible:

"For no one can lay any foundation other than the one already laid, which is Jesus Christ. ¹² If anyone builds on this foundation using gold, silver, costly stones, wood, hay or straw, ¹³ <u>their work will be shown for what it is, because the Day will bring it to light. It will be revealed with fire, and the fire will test the quality of each person's work. ¹⁴ If what has been built survives, the builder will receive a reward. ¹⁵ If it is burned up, the builder will suffer loss but yet will be saved—even though only as one escaping through the flames</u>."
(1 Corinthians 3:11-15)

Every pastor is eager to get his church's attendance to grow and the total number of the membership at his church to multiply; but few are actually concerned with the quality of spirituality the members of their congregations go home with and practice through the week.

A true measure of any church's effectiveness is the improvements in the godly living of members of its congregation. Even if only one soul continues to improve in their godliness from one week to another, that is a serious gain for the kingdom of God; and a good reflection on God and His promises in the Bible.

The spiritual improvement in that one person's life guarantees that Jesus Christ will be present in the church services Sunday after Sunday, because the church is helping its members improve their spirituality. The spiritual betterment of the congregation counts much more to God than the size of the congregation. A congregation that multiplies by leaps and bounds but produces no fruit of the Spirit from one Sunday to another does not impress God.

On the other hand, a congregation that has only a few members but does wonders in the spiritual growth of those few members receives the assurance of God's presence among them every time they meet. All that matters is that whenever a group of people meet in the name of God, they must interact with one another in truth and in the Spirit.

The quality of their interaction then brings the Holy Spirit into their midst. God is not interested in a large number of people getting together regularly to promote their personal agenda, disguised as a service to God and a fellowship with one another. The church must never be used as a cover for financial schemes and personal atrocities.

The church of Jesus Christ is not the equivalence of the old Temple of God in Jerusalem. The church of God is the equivalence of the ancient Israel as a society, which the temple is only a part of; and not all of.

The totality of Christianity constitutes the people of God which is the equivalence of the Israelites of the old. And because God gave the ancient Israel a real economy, God has also given the church a real economy.

The church has only failed to recognize this fact and has instead subscribed to the world's economy; while the ancient Israel on the other hand contributed valuable ideas to the world economy; and the current nation of Israel continues to do so today.

The greatest reason why ancient Israel succeeded in operating according to God's economy and the church has failed to operate according the kingdom economy is as follows:

Whereas ancient Israel recognized itself as one people in God, the church so far has failed to recognize itself as one people in God. The church believes itself to be one people only in principle. But it has failed to transform that into a reality.

The church continues to look at its stature in the world as being more important than the true blending of all of its members as people of one family—the family of God which is like every other real family in the world. It is the realness of this family—which Jesus Christ established and handed over to the apostles, under the leadership of Apostle Peter—that convinced the first century Christians to pursue a common life for all of their numbers. They lived together when possible, shared together and suffered together.

They constantly encouraged and strengthened one another as members of one family; and not as independent families that share common habitat. They looked after one another and were passionate about their numbers. They were emotionally vested in the lives of one another.

These days, you can sit in a church and the person sitting next to you does not even want to talk to you for whatever their reason might be. How can people who come together weekly but have no interest in even knowing the other members of the congregation have a shared spiritual life?

We pretend that we do, but the reality is starkly contrasting to our assumption. People, within a community, bond more readily over common civic issues within their community than believers bond with the believers that attend their church. Church or religious life has come to be far more inferior to our civic lives. The things of God have become secondary to everything else in our lives.

For most of us, religious life has become like going to a movie. You go in and attend service and rush out of the door as soon as the service is over. We attend these services not because they satisfy anything in our lives, or fill us with spirituality. But simply because we count it as an obligation we must fulfill because we have become convinced that without doing that we will run the risk of going to hell. So, going to church and doing anything with others at the church has become just another civic duty. Therefore, how can such an experience make anyone bond with anyone else in Christ?

Everyone is interested in avoiding hell but few are really interested in serving our brothers which translates into serving God. We look at anybody in our community who does not look exactly like us as if they would attack us any second and cause harm to us. No matter what good gestures they make to us, we are never convinced they would not do something sinister to us. At the same time, we walk into real danger in the hands of some people we trust, because they look like us and have the kind of lifestyle which we approve. And out of fear of losing financial support, the church avoids preaching against such discriminatory behaviors.

The church pretends that we are all coming together in Christ as required by the Bible because the church has no real effect on any of us but is rather at our mercy for its financial survival. Well, a church which is afraid of losing revenue is like a dog with no teeth and has no effectiveness in leading the congregation to godly living.

Instead it operates under the dictates of its most affluent and most vocal members. It has become another outlet for the rich and powerful to extend their influence over the people; instead of being a place that rescues the powerless from the powers of exploitation and subjugation.

The church has an anemic economy because the church does not recognize its people as one people—with each member equally vested in the family of God. The church's focus on the viability and stature of the religious institution is more important to the church leadership than true spiritual investment in the believers the church serves. The church does not see itself as being charged with providing for anyone but itself.

The church is rather interested in operating as the Temple of God; rather than operate as the kingdom of God, which it is. And every kingdom has a true economy, therefore, the church needs to embrace its status as a kingdom and develop the economy God has favored it with. That way the church will become of service to all of its members and help eradicate poverty from within the church of God.

God is not interested in our charity to one another. God is interested in believers making real investments into other believers' lives; just the same way people who are members of the same family do.

God is interested in the church leadership encouraging and arranging for believers with business and technical expertise to recruit other believers in their expertise to help boost the kingdom economy.

The tithes and offerings received by the church are intended to first and foremost serve the needs of believers within the church. This is the most effective way to spread the gospel. The good deeds a church does for its members carries far and wide and attracts non-believers to become believers.

This is more effective that getting behind the microphone every day and talking. There is a common saying that action speaks louder than words. And Jesus Christ commanded believers everywhere to become salt and light to the world. This translates to believers and the church becoming a source of inspiration in every way to the world around them.

We should attract the world through our good deeds and not through empty words. God puts His power behind our deeds, not our words. If the church focuses on letting its actions speak for it, the church would become the economy that Jesus Christ designed it to be.

When the word got out that Jesus Christ fed the followers abundantly, more people flocked around Him to get their own fill. And to that Jesus Christ said:

"Very truly I tell you, you are looking for me, not because you saw the signs I performed but because you ate the loaves and had your fill. ^{27} Do not work for food that spoils, but for food that endures to eternal life, which the Son of Man will give you. For on him God the Father has placed his seal of approval." (John 6:26-27)

Stories about the church doing things for the

economic relief of the believers would make the biggest draw for the church. There will be more stories of miracles coming out of that church every week because all the believers would go around doing what they do believing that Jesus Christ will come true for them as He promised in the Bible. And He would, because more believers would be operating with common spirit and in obedience to God. Obedience to all of God's commands is the truest way of worshipping God.

CHAPTER SEVEN

The Power of Grace

A number of Christian preachers teach Grace as if Grace is some New Testament thing from God. That idea is in error. Jesus Christ did not originate in the four gospels of Matthew, Mark, Like and John. Jesus Christ was with God before the beginning of time and before the rest of God's natural creations as captured in Genesis Chapter One.

That is why the gospel of John started with:

"In the beginning was the Word, and the Word was with God, and the Word was God. ² He was with God in the beginning. ³ Through him all things were made; without him nothing was made that has been made. ⁴ **In him was life***, and that life was the light of all mankind. ⁵ The light shines in the darkness, and the darkness has not overcome it.*

⁶ There was a man sent from God whose name was John. ⁷ He came as a witness to testify concerning that light, so that through him all might believe. ⁸ He himself was not the light; he came only as a witness to the light.

⁹ The true light that gives light to everyone was coming into the world. ¹⁰ He was in the world, and though **the world was made through him***, the world did not recognize him. ¹¹ He came to that which was his own, but his own did not receive him. ¹² Yet to all who did receive him, to those who believed in his name, he gave the right to become children of God— ¹³ children born not of natural descent, nor of human decision or a husband's will, but born of God.*

¹⁴ The Word became flesh and made his dwelling among us. We have seen his glory, <u>the glory of the one and only Son, who came from the Father, full of grace and truth</u>." (John 1:1-14)

Grace has always been around as long as God has been around. *Grace* is the Spirit of God—*the substance of all creations*. Grace is Jesus Christ, the Son of God, *in* whom, *through* whom, and *for* whom, everything was made *(Colossians 1:15-20)*. This is why Jesus Christ said to the Jews:

"You study the Scriptures diligently because you think that in them you have eternal life. <u>These are the very Scriptures that testify about me, ⁴⁰ yet you refuse to come to me to have life.</u>" (John 5:39-40)

The whole Bible and everything faith accomplishes in the Bible demonstrates the power of Grace:

It was Grace that fulfilled every word and desire of God when God created the earth and the universe *(Genesis 1:1-31)*.

It was Grace that allowed Adam to name all the

animals God brought to him to give names to *(Genesis 2:19)*.

It was Grace that allowed Adam to profess that Eve was taken from him *(Genesis 2:23)*; when in reality Adam had no knowledge of what had happened to him because he was under anesthesia, and so did not personally witness what transpired.

It was Grace that led Eve to proclaim that God had helped her bring forth a man—a baby *(Genesis 4:1)*.

It was also Grace that led Abel to figure out what to sacrifice to God because Abel had a good heart and wanted to please God *(Genesis 4:3-5)*. Abel's desire to please God brought him more Grace and allowed him to accomplish his desire to please God. That is why the Bible tells us to ask and it will be done to us. Your desire to do good brings you the Grace you need to do good!

It was Grace that set Enoch apart from the rest of the people of his time, and made Enoch a prophet among them. After Enoch was done with his work on earth, Grace took him home to God, without him having die first *(Genesis 5:21)*.

It was Grace that familiarized Noah with the voice of God and allowed Noah to understand God and follow God's instructions diligently for more than 120 years; when the world around him was sinful and devoid of the Spirit of God *(Genesis 6:5-22)*.

It was Grace that safeguarded the Noah's ark from the destructive forces of the storm and the deluge that destroyed the entire world *(Genesis chapters 6-9)*

It was Grace that made Abram obey the voice of God to go to a land God would show him, which would be given to his descendants as an inheritance *(Genesis 12:1-9)*.

It was Grace that kept Sarah on Abraham's side; ministering to his needs and putting herself in apparent danger to save her husband's life, time and time again *(Genesis 12:11-20) (Genesis 20:1-18)*.

It was Grace that made Abraham pursue the rogue nations that banded together and attached Sodom, taking his beloved nephew, Lot, and his family captives. Through the power of Grace, Abraham recovered Lot and his family and restored them to their rightful place with only a handful of men *(Genesis 14:1-24)*.

It was Grace that prompted Sarah to make a selfless sacrifice by giving her maid to her husband in an attempt to help fulfill the promise God made to her husband *(Genesis 16:1-15)*.

It was also Grace that rewarded Sarah by giving her the fruit of the womb when she was way past the age of childbearing; as a reward for her faith in God and her loyalty to her husband *(Genesis 17:15-*

16). Grace made Sarah a virtuous woman!

It was Grace that saved Lot and his family from the destruction of Sodom by fire. And disobedience and unfaithfulness caused Lot's wife the same life she was spared form the fire, causing Lot to live in fear for the rest of his life. But Grace maintained Lot's lineage through what was naturally unlawful relations, because Lot was blameless. *(Genesis chapters 18 & 19)*

It was Grace that saved Hagar and Ishmael when they were sent away by Abraham due to Sarah's distress, and made Ishmael prosperous *(Genesis 21:1-21)*.

It was Grace that made Isaac stay back on the land instead of going to Egypt like his father, Abraham, to avert the ravages of a famine. Grace made Isaac prosper greatly when others diminished immensely, making Isaac the envy of all the people. *(Genesis 26:1-32)*

It was Grace that brought Rebecca the news about the fate of her twin babies *(Genesis 25:21-23)*.

And it was Grace that caused Rebecca to favor the younger twin over the older one, in spite of the inherent danger to her life and the disintegration of her family *(Genesis 27:1-46)*.

It was Grace that kept Jacob throughout his trials

after his deceit of his twin brother and was on the lam. Jacob prospered, in spite of his father in-laws deceptive and dishonest treatment of him. He returned home to his own land after twenty years in exile. And grace placated his brother Esau and set Jacob free from guilt and sin. *(Genesis Chapters 28-31) (Deuteronomy 32:9-14)*

It was Grace that preserved the life of Joseph through his captivity in Egypt and elevated him to the highest position in the land, in a matter of thirteen years. Grace kept him from soiling his hands by involving himself with his master's wife or other wicked things. And Grace made Him rule justly and saved all the people of Egypt from a devastating famine. *(Genesis Chapters 39-48)*

It was Grace that divided the waters of the Red Sea at the postulation of Moses to allow the people of Israel to safely cross to the other side and escape the pursuing mighty Egyptian army. Grace let all Israel through, and devastated the entire Egyptian army who went into the sea in pursuit of the Israelites. *(Exodus 14:10-31)*

It was Grace that turned the bitter water of Marah in the desert to drinkable water. *(Exodus 15:22-27)*

It was Grace that started a spring of drinking water out of the rock at Kadesh in the Desert of Zin, through Moses's postulation. Moses in his anxiety had done something counter to what God commanded him to do, yet Grace gave the people of God the water they desired. *(Numbers 20:8-13)*

It was Grace that caused the snakebites sustained by the Israelites, as they went through the wilderness, to heal just by each bitten Israelites looking up at the raised statute of the snake as commanded by God. *(Numbers 21:4-9)*

It was Grace that healed the infectious skin diseases that infect the skins of the Israelites as they journeyed through the wilderness to the Promised Land. Just by isolating the infected person outside the camp for seven days and cleansing their infected body as commanded by God, they returned to their place in the community good as new. *(Deuteronomy 24:8-9)*

Bacterial infection, fungal infection, even viral infection; it did not matter. Simply by complying with God's commands, the infection completed cleared within seven days, and the infected persons returned to the community without posing any threats to the community at large.

It was not the concoction that cures the infected person. It was the Grace of God inside the concoction that caused the necessary cleansing and eliminated the various diseases completely.

God knew that the world would only get more sinful; and the human life, more desperate. So, through the cures God taught Moses and Aaron in the wilderness, God was setting precedence for humanity on methods of controlling and curing diseases.

It is not our concoctions that cure our diseases today. It is the Grace of God in the concoctions that cure all our diseases and restores our health to normal. Everything in the world and the universe runs on Grace! Without Grace, nothing exists because Grace is God the Father, God the Son and God the Holy Spirit.

That mankind did not discover germs and their handling through the expert methods God taught these two pioneers, demonstrates man's complete ignorance of God, and shows our disdain for Him.

The rest of the Bible is filled with instances of the power of Grace. Grace increases as faith in God and His Christ increases. Faith always summons Grace; and Faith concentrates Grace to levels unthinkable.

CHAPTER EIGHT

The World has never seen a Single Religious War

There really has been a religious war in the world. Even the infamous Christian-Muslim wars were fought more for economic and political dominance, than they were fought for religious dominance.

Religion was simply a cover. The real reasons behind the so-called religious wars in our history books were politics, economics, ideologies; and the desire to dominate the world.

Every dominant power that overtakes another; imposes its religion and ideologies on the kingdom it dominates. That has always been the case throughout human history.

That a dominant power forces its religion on the kingdom it dominates does not make the dominance a religious cause. Religions have never really been at war with each other, because religions, characteristically, appeal quietly to the soul, instead of antagonizing the soul.

Religions tend to placate—not aggravate! People who seek religion are looking to quell the turmoil in their lives.

That the world has got the nomenclatures so confused as to include occults as religions does not mean that anyone should confuse religion with occult. Seeking dark powers is the domain of the devil and has never provided any religious benefits to anyone. Religion serves true human needs.

Politics, on the other hand, uses all the force it can muster, to intimidate and control whoever it comes against. Politics is the brute that asserts itself over all human society and works to control totally everybody and everything under its commands.

Politics hides under religion to exert its authority and power so that it may achieve its objectives. Politics is the beast of Revelation who exploits humanity to elevate itself above everything that is God.

The problem facing humanity and the world is not religion. Rather, it is the gruesome and beastly Politics! Politics despises religion. Yet Politics, in its parasitic nature, extravagantly taps into whatever benefits it could derive from religion.

Every religion in the world provides some spiritual benefits to its practitioners. Otherwise, no one would subscribe to the religion. Therefore, a true religious war will be a war between two religions to demonstrate which religion would provide the most spiritual benefits to its practitioners.

Such would be a war to outperform the other religion in the provision of spiritual benefits to the people of the community. Such a war would provide no benefits to the devil. Therefore, why would the devil bring people together to fight such a war? Wars are caused by selfish ambitions; stirred up in people by the devil and wicked desires. Wars are never caused by the desire to do "good" in society!

People voluntarily become involved in religions to seek spiritual benefits; and that is a noble ambition. Most people that seek religion are looking for solutions to counter the effects of the devil in their lives; not to amplify them. It is through this desire to find spiritual solutions that people come to God, through the power of Grace *(Acts 17:27)*. Therefore, the desire for spirituality can never be at war with itself—or with God, for that matter.

God is too great to be in competition with anything which He created. God is not into religion—even though He understands the importance of religion in our lives—because we often confuse religion with what is really important. God is all about godliness!

If God is into religion, no other religion in the world

would see the light of the day against God's religion—not even by a long shot. Every other religion would be dead as soon as it comes into existence because it would have nothing to stand on. Mankind always does the wrong thing with religion: Sooner or later, we replace the real substance of the religion with ordinary human traditions.

Evil does not compete against Grace because evil does not stand any chance whatsoever against Grace—just in the same way darkness does not stand a chance against light. Grace dissipates and neutralizes evil; the same way light swallows up darkness. So if God is into religion, evil cannot withstand that religion.

Religion crops up around the work of Grace, because Grace transcends religion. God created all mankind in Grace and through Grace. And God controls all mankind through Grace. Through Grace, God meets all human needs for all of mankind, regardless of language or creed. Even for those who have no religion, Grace works their entire lives. Therefore, they must abide by the commands of God or ruin their lives.

Unfortunately, after religion is started by Grace, religion then tries to impose itself over Grace. And before you know it, religion is all you can see, because subsequent generations of the practitioners continue to infuse their own personal ideas into the religion in an attempt to either modernize the religion, or make it better.

If God were into religion, that religion will be handed to human beings already perfected. Yet human beings will try to improve on it and mess it up.

When people categorize Christianity as a religion, they minimize the power of Grace. Religion is the worship of some deity. And the people in a religion do not necessarily have to know what they are worshipping, but they worship nonetheless.

The Bible is not a book of codes about membership to an exclusive organization of worshippers called the Christians. The Bible is the Book of life for all mankind; wherever they find themselves in the world.

The Bible is not designed by God for only those who believe in Him and/or His Son Jesus Christ. The Bible is designed by God for all human beings everywhere in the world; to lead every human being to find God. That is why the Bible says:

"Paul then stood up in the meeting of the Areopagus and said: "People of Athens! I see that in every way <u>you are very religious</u>. ²³ For as I walked around and looked carefully at your objects of worship, I even found an altar with this inscription: TO AN UNKNOWN GOD. So you are ignorant of the very thing you

worship—and this is what I am going to proclaim to you.

²⁴ "The God who made the world and everything in it is the Lord of heaven and earth and does not live in temples built by human hands. ²⁵ And he is not served by human hands, as if he needed anything. Rather, he himself gives everyone life and breath and everything else. ²⁶ From one man he made all the nations, that they should inhabit the whole earth; and he marked out their appointed times in history and the boundaries of their lands. ²⁷ God did this so that they would seek him and perhaps reach out for him and find him, though he is not far from any one of us. ²⁸ 'For in him we live and move and have our being.' As some of your own poets have said, 'We are his offspring.'

²⁹ "Therefore since we are God's offspring, we should not think that the divine being is like gold or silver or stone—an image made by human design and skill. ³⁰ In the past God overlooked such ignorance, but now he commands all people everywhere to repent. ³¹ For he has set a day when he will judge the world with justice by the man he has appointed. He has given proof of this to everyone by raising him from the dead.'" (Acts 17:22-31)

God has overlooked people of the world worshipping different things, most of which they did not know, and has finally sent His Son Jesus Christ to die for the sins of the world so that whoever believes in that sacrifice He made with His own life will be saved and have the life that is truly life.

Christ's sacrifice is not a religious thing intended for only people who share the label of the Christian religion. Christ's sacrifice is life itself! And that life is given as a gift to all mankind.

And God did not intend for any human being to be left out of that gift. That is why all that is required for any human being to receive that special gift of life is faith in the one who made the sacrifice—the one who was appointed by God to be the channel through which all human beings will be reconciled to God.

To understand that God counts only godliness, and not religion, look at the following passage. This is right after Apostle Paul made the declaration in Athens about God now demanding that all human beings desist from worshipping the things they do not know, and come to know Jesus Christ, and partake of His great sacrifice for all mankind:

"After this, Paul left Athens and went to Corinth. ² There he met a Jew named Aquila, a native of Pontus, who had recently come from Italy with his wife Priscilla, because Claudius had ordered all Jews to leave Rome. Paul went to see them, ³ and because he was a tentmaker as they were, he stayed and worked with them. ⁴ Every Sabbath he reasoned in the synagogue, trying to persuade Jews and Greeks.

⁵ When Silas and Timothy came from Macedonia, Paul devoted himself exclusively to preaching, testifying to the Jews that Jesus was the Messiah. ⁶ <u>But when they opposed Paul and became abusive, he shook out his clothes in protest and said to them, "Your blood be on your own heads! I am innocent of it. From now on I will go to the Gentiles."</u>

⁷ Then Paul left the synagogue and went next door to the house of Titius Justus, a worshiper of God. ⁸ Crispus, the synagogue leader, and his entire household believed in the Lord; and many of the Corinthians who heard Paul believed and were baptized.

⁹ One night the Lord spoke to Paul in a vision: "Do not be afraid; keep on speaking, do not be silent. ¹⁰ For I am with you, and no one is going to attack and harm you, because I have many people in this city." ¹¹ So Paul stayed in Corinth for a year and a half, teaching them the word of God.

¹² While Gallio was proconsul of Achaia, <u>the Jews of Corinth made a united attack on Paul and brought him to the place of judgment.</u> ¹³ "This man," they charged, "is persuading the people to worship God in ways contrary to the law."

*¹⁴ Just as Paul was about to speak, Gallio said to them, "**If you Jews** were making a complaint about some misdemeanor or serious crime, it would be reasonable for me to listen to you. ¹⁵ But* *since it involves questions about words and names and your own law—settle the matter yourselves. I will not be a judge of such things." ¹⁶ So he drove them off. ¹⁷ Then the crowd there turned on Sosthenes the synagogue leader and beat him in front of the proconsul; and Gallio showed no concern whatever." (Acts 18:1-17)*

Apostle Paul, in obedience to the commands of Jesus Christ, preached to the Jews first, at their synagogue. But when they opposed him and became abusive to Him, Apostle Paul <u>"shook out his clothes in protest and said to them, "Your blood be on your own heads! I am innocent of it. From now on I will go to the Gentiles.""</u> And he did go to the Gentiles.

Apostle Paul was preaching to the religious people who were supposedly worshipping God, yet most of them rejected the true word of God which he preached to them. Those believers had religion, but they did not have God! They flocked to the house of God as stipulated in the Book of law of God, yet they were worshipping something else—not God. Otherwise, they would have recognized and accepted the truth of God Apostle Paul was preaching to them.

They had religion. And their religion had become more important to them than the truth of God, which the religion was designed to foster and promote. As a result, they failed to recognize the truth when the truth was proclaimed to them; prompting the apostle that brought the *"good news"* to shake out his clothes in protest and move his venue outside the established house of God.

Their religion rose from the word of God. But over time, their religion became more important to their lives than the truth of God. Somehow, they have managed to leave the truth of God behind, and went on with a religion that has become nothing more than a civil code that can be effectively enforced to keep the people in line.

This is why Jesus Christ Himself chided the leaders of the religious order, telling them:

""Isaiah was right when he prophesied about you hypocrites; as it is written:

*"'These people honor me with their lips,
 but their hearts are far from me.*
*⁷ They worship me in vain;
 their teachings are merely human rules.'*

⁸ You have let go of the commands of God and are holding on to human traditions."

⁹ And he continued, "You have a fine way of setting aside the commands of God in order to observe your own traditions! ¹⁰ For Moses said, 'Honor your father and mother,' and, 'Anyone who curses their father or mother is to be put to death.' ¹¹ But you say that if anyone declares that what might have been used to help their father or mother is Corban (that is, devoted to God)— ¹² then you no longer let them do anything for their father or mother. ¹³ Thus you nullify the word of God by your tradition that you have handed down. And you do many things like that."

¹⁴ Again Jesus called the crowd to him and said, "Listen to me, everyone, and understand this. ¹⁵ <u>Nothing outside a person can defile them by going into them. Rather, it is what comes out of a person that defiles them.</u>"" (Mark 7:6-15)

The Lord came to Apostle Paul in a dream after the apostle had moved his preaching from the established house of God in Corinth, into a person's house. The Lord assured the apostle of his safety and continued guidance; and told him that He—the Lord—had many people in the city of Corinth who would come to the apostle's aid whenever the need arose. Who were these *"many people"* the Lord just spoke to the apostle about?

Were they Jews or were they Gentiles? Or were they a mixture of Jews and Gentiles? The answer is obviously a mixture of Jews and Gentiles: Jews who would stand with the apostle on the truth of the gospel he was preaching; and Gentiles who would provide necessary cover for the apostle to protect the apostle from the vile and violent Jewish religious leaders in the city of Corinth.

In addition to counting on the righteousness and support of discerning Jews in the city of Corinth, God was also counting on the godliness of some Gentiles in the city to provide necessary protection from persecution by the violent religious people to His servant Apostle Paul as the apostle presented the gospel of Grace in the city of Corinth.

According to *Acts 18:8*, Crispus, the synagogue leader, and his entire household, believed the truth Apostle Paul was preaching, and were baptized. Many of the Corinthians who heard Paul also believed and were also baptized. But since the opposition to the apostle's teaching was great, the apostle had to move his teaching to a nearby residence and continued from there. These are the *many people* the Lord spoke to the apostle about.

These religious people who have obviously walked away from the real substance of their religion—Grace—took the apostle to the law enforcement and pressed charges, in an attempt to silent him altogether.

But, like the Lord had revealed to the apostle, the Gentile in charge—*Gallio, proconsul of Achaia*—had the wisdom to absolve himself from hearing the case, citing that the charges brought against the apostle stemmed from religious interpretations; and not civil disobedience.

The decision of a wise and conscientious governor now prevented the fanatical religious leaders from dealing a death blow to a true messenger of the gospel; and allowed the gospel to get to the people, as intended by God.

A religious order, which rose from the truth of God but has become devoid of Grace, pitched itself against the truth of God, and lost; once again, demonstrating the power of Grace over human ideologies and human establishments and their wicked intents.

The Jewish religion rose from the truth of God. Yet throughout the New Testament, the leaders of the religion stood opposed to the truth of God that came from Grace. The Christian religion also rose from the truth of God. Yet, the Christians have distorted the truth of God more than people outside the religion; not to mention the all-out effort made by Christian leadership in the middle ages to make the Jewish religion extinct.

God and the Bible are not about religion. They are about the truth and life—both of which are the rights of every human being that exists anywhere in the world, regardless of language or creed. Like Jesus Christ said:

"Nothing outside a person can defile them by going into them. Rather, it is what comes out of a person that defiles them."" *(Mark 7:15)*

Our thoughts, speech and actions have the power to either endear us to God or remove us from His presence and His inheritance. Our ideologies, if not wholly supported by the truth of God, put us on a collision course with destiny. It is not our religion that matters to God. It is our godliness that is all the rage.

It is better not to have a religion and yet care highly about the welfare of your fellow human beings, than to be highly religious and tune out human grief and sufferings. In essence, it is better to be godly and have no religion than to be religious and be disobedient to God. Your religiosity must not outpace your obedience to all of God's commands and directions.

Rather, your conformity to godly living must set the course for your religiosity—just like it did for Cornelius the centurion, whose godliness compelled God to send Apostle Peter to bring Cornelius and his entire household to Jesus Christ. That is why the Bible says:

"Those who consider themselves religious and yet do not keep a tight rein on their tongues deceive themselves, and their religion is worthless. Religion that God our Father accepts as pure and faultless is this: to look after orphans and widows in their distress and to keep oneself from being polluted by the world."
(James 1:26-27)

God never fails to reward godly acts!

Is religion therefore unimportant? That is far from the truth. That people of every generation have tainted and misused religion does not diminish the importance of religion in the human lives. Religion has always been about spirituality. Religion is belief in powers unseen—powers above human powers. Religion is always a good thing because religion discourages people from perpetrating absolute evil.

All religions have the power to mislead because no religion is insulated from human meddling and deceit. And while all religions are set up to offer some spiritual benefits to the people in the religions, many religions lack essential truths of God; especially the truths that pertain to creation, sin, remediation of sins and salvation.

That is why no one should allow their religion to block them from learning the truth of God as it pertains to those areas; because there are grave consequences for being ignorant of the truth. In the end, it is not what we choose to accept as the truth that matter. It is our willingness to learn the truths and embrace all of them for our benefits that really matter.

That is the chief benefit of knowledge. That a religion is indigenous to your country does not mean it is the right religion for you. Going after the truth may prove costly to your life; but that is a price that is what paying. Examine the truths laid out in my many book and other well-intended books and decide for yourself how best to seek God's truth and salvation. Your life depends on it, and time is of the essence!

For the Christians who are worried about the antichrist, what truly threatens Christianity, and the belief in God in general, is non-spirituality. Anyone who honestly pursues spirituality respects others who pursue spirituality—aside from the deadly imitators who practice Satanism as their form of spirituality. No belief in a higher power that brings good to the people poses any real threat to the belief in the one God of the earth and the universe.

The real threat to the Christian faith, and belief in God, is secularism which seeks to entirely exclude spirituality from the human society. The belief that the human beings are the creators, and the masters, of their own destiny is really detrimental to humankind. That is the true antichrist that will turn the world on its head. The antichrist is not another current world religion! That is what the real antichrist wants everybody to believe, so he could continue to fly under the cover.

Most people in the different religions of the world care about the betterment of their own welfare more than they worry about other religions of the world overtaking theirs. Since different religions are predominantly practiced in different geographies of the world, political and ideological ambitions are often masked under religion, thereby making political and economic bickering seem like religious antagonism.

The world scientists, who were once Christians, but have renounced their Christian faith and now place their trusts entirely on science—or are nominal Christians and believe in the supremacy of science over everything else in the human life—constitute the real threat to the Christian faith; and to any other faith that believes in the one God of the earth and the universe.

That is why the current belief of many Christian believers and their teachers that a different religion will, sometime in the future, overpower the

Christians and force them to renounce their God is misguided.

The only documented successful godly government in the Bible was the government of a Gentile nation—the ancient Egypt—with a wise and discerning leader, Joseph, at the helm of that government.

Joseph, as the sole ruler over that government, continued to give food and other allowance to the heathen priests in that government as was the practice, and God did not object to that. God's central concern was that all the people of the kingdom be provided for; and they were in all intent and purposes.

Religiosity was never called into question in all the years Joseph provided godly leadership to that kingdom. The equitable maintenance of all human lives was the singular ambition of that government. And the government met that objective in a spectacular way:

Systematically, God caused the entire wealth of that nation to be accumulated under the wise supervision of His dedicated servant, Joseph; to ensure that all the needs of all the citizens of that kingdom were met—once again demonstrating the constant theme in the Bible that it is the express will of God for all the citizens of the world to be sufficiently fed, clothed and housed. Taking sufficient care of the people of every nation is more important to God than any religious ideals.

Any people, anywhere in the world, who through godly leadership receive adequate care and consideration for their welfare, would more readily ascribe to the commands of God written in the Bible. The constant pressures to make the ends meet always get the people of the world communities running after whatever promises to help alleviate their sufferings. Any religion that caters to the welfare of any people captures the loyalty of that people.

And there is nothing in the world that delivers as much as it promises more than the Bible does. Yet, the way many Christians proclaim the things of the Bible, but live in a way that is contrary to the Bible, casts doubts in other peoples' minds about the Christian faith. It is that attitude that makes the Christian faith a mere religion in the eyes of other people—at par with the religions of the world. That attitude also minimizes the rewards those people receive for their faith in God and the Bible.

The Christians who reduce the Christian faith to a mere religion are ignorant of God and the power of Grace. Their practice of Christianity is a disservice to the peoples of the world who thirst for the truth of God and life. The Bible does not promote the practice of the Christian faith as an exclusive club of entitlement. The Bible is the instrument God designed to bring all human beings to the truth, and to salvation.

I repeat: God and the Bible are not about religion. That is why the Jewish religion and the Christian religion still differ on Grace. God and His Bible are about the truth and life—both of which are the rights of every human being that exists anywhere in the world, regardless of language or creed. If you are a human being, you have the right to life and to the truth, but you must be willing to seek them as commanded by God; or both will elude you.

But you will not know what God commands unless you read the Bible. So read the Bible, even if you are not a Christian. Read the Bible in search of the truth of God, as it pertains to creation, sin, remediation of sins and salvation. All of us have read some books that have nothing to do with our religion.

Adding the Bible to that list can become a life-saver for you. Therefore, do not hesitate to read the Bible and begin the journey to your salvation. We read books on economics, sciences, astronomy and even astrology.

Incidentally, Life is the Truth who is Grace—who is the person of Jesus Christ—the only begotten Son of God *(John 10:36)*, who declared in the gospel:

"I and the father are one" *(John 10:30)*.

And to shed more light on this great revelation, Jesus added:

"Do not believe me unless I do the works of my Father. ^{38} But if I do them, even though you do not believe me,

believe the works, ***that you may know and understand that the Father is in me, and I in the Father.****" (John 10:37-38)*

Nobody's current religion makes them ineligible to share in the Truth and the Life who is Jesus Christ—in much the same way a person's religion does not disqualify them from seeking scientific facts.

Some are allowing their religious affiliations to preclude them from the gift of life and truth that is Jesus Christ. Do not be deceived! All anybody needs to do to qualify to receive the gift of truth and life is to show the desire and willingness to receive the gift; and to have faith in both the gift and the one giving the gift.

When you are born a human being, you were given life by God. In other words, being born human automatically makes you eligible to hear the truth of God that is posited for mankind in the Bible and believe in the sacrifice made on your behalf by our Lord Jesus Christ. Listen to Jesus Christ Himself, as He takes you through the whole truth about your rights to partake of His great gift of life:

""Very truly I tell you Pharisees, ***anyone*** *who does not enter the sheep pen by the gate, but climbs in by some other way, is a thief and a robber.* ² *The one who enters by the gate is the shepherd of the sheep.* ³ *The gatekeeper opens the gate for him, and the sheep listen to his voice. He calls his own sheep by name and leads them out.* ⁴ *When he has brought out all his own, he goes on ahead of them, and his sheep follow him*

because they know his voice. ⁵ But they will never follow a stranger; in fact, they will run away from him because they do not recognize a stranger's voice." ⁶ Jesus used this figure of speech, but the Pharisees did not understand what he was telling them.

⁷ Therefore Jesus said again, "Very truly I tell you, I am the gate for the sheep. ⁸ All who have come before me are thieves and robbers, but the sheep have not listened to them. ⁹ <u>**I am the gate; whoever enters through me will be saved.**</u> *They will come in and go out, and find pasture. ¹⁰ The thief comes only to steal and kill and destroy;* <u>**I have come that they may have life, and have it to the full.**</u>

¹¹ "<u>**I am the good shepherd. The good shepherd lays down his life for the sheep**</u>*. ¹² The hired hand is not the shepherd and does not own the sheep. So when he sees the wolf coming, he abandons the sheep and runs away. Then the wolf attacks the flock and scatters it. ¹³ The man runs away because he is a hired hand and cares nothing for the sheep.*

¹⁴ "<u>*I am the good shepherd; I know my sheep and my sheep know me—* ¹⁵ *just as the Father knows me and I know the Father*</u>*—and I lay down my life for the sheep. ¹⁶* <u>**I have other sheep that are not of this sheep pen. I must bring them also. They too will listen to my voice, and there shall be one flock and one shepherd.**</u> *¹⁷ The reason my Father loves me is that I lay down my life—only to take it up again. ¹⁸ No one takes it from me, but I lay it down of my own accord. I have authority to lay it down and authority to take it up again. This command I received from my Father."*

¹⁹ The Jews who heard these words were again divided. ²⁰ Many of them said, "He is demon-possessed and raving mad. Why listen to him?"

²¹ But others said, "These are not the sayings of a man possessed by a demon. Can a demon open the eyes of the blind?"

²² Then came the Festival of Dedication at Jerusalem. It was winter, ²³ and Jesus was in the temple courts walking in Solomon's Colonnade. ²⁴ The Jews who were there gathered around him, saying, "How long will you keep us in suspense? If you are the Messiah, tell us plainly."

²⁵ Jesus answered, "I did tell you, but you do not believe. The works I do in my Father's name testify about me, ²⁶ but you do not believe because you are not my sheep. ²⁷ My sheep listen to my voice; I know them, and they follow me. ²⁸ I give them eternal life, and they shall never perish; no one will snatch them out of my hand. ²⁹ My Father, who has given them to me, is greater than all; no one can snatch them out of my Father's hand. ³⁰ I and the Father are one."

³¹ Again his Jewish opponents picked up stones to stone him, ³² but Jesus said to them, "I have shown you many good works from the Father. For which of these do you stone me?"

³³ "We are not stoning you for any good work," they replied, "but for blasphemy, because you, a mere man, claim to be God."

³⁴ Jesus answered them, "Is it not written in your Law, 'I have said you are "gods"'? ³⁵ If he called them *'gods,' to whom the word of God came—and Scripture cannot be set aside— *[36]* what about the one whom the Father set apart as his very own and sent into the world? Why then do you accuse me of blasphemy because I said, 'I am God's Son'? *[37]* Do not believe me unless I do the works of my Father. *[38]* But if I do them, even though you do not believe me, believe the works, that you may know and understand that the Father is in me, and I in the Father." *[39]* Again they tried to seize him, but he escaped their grasp.*

[40] *Then Jesus went back across the Jordan to the place where John had been baptizing in the early days. There he stayed, *[41]* *and many people came to him. They said, "Though John never performed a sign, all that John said about this man was true." *[42]* *And in that place many believed in Jesus." (John 10:1-38)*

Jesus Christ was talking about those of us who are not of Jewish descent, who nonetheless heard the gospel and believed and placed our trust in Him, when He said:

"I have other sheep that are not of this sheep pen. I must bring them also. They too will listen to my voice, and there shall be one flock and one shepherd."

So no one's current religion, or lack thereof, should prevent them from responding to the voice of the great Shepherd of life and truth. It is not only those outside the Christian faith that allow religion to blind them from the truth of God. Majority of the people within the Christian religion see their religion as a substitute to obeying all the commands of Jesus Christ. And it is to this group of Christians that Jesus Christ said the following:

""Not everyone who says to me, 'Lord, Lord,' will enter the kingdom of heaven, but only the one who does the will of my Father who is in heaven. ²² Many will say to me on that day, 'Lord, Lord, did we not prophesy in your name and in your name drive out demons and in your name perform many miracles?' ²³ **Then I will tell them plainly, 'I never knew you. Away from me, you evildoers!'**

²⁴ "<u>Therefore everyone who hears these words of mine and puts them into practice is like a wise man who built his house on the rock.</u> ²⁵ The rain came down, the streams rose, and the winds blew and beat against that house; yet it did not fall, because it had its foundation on the rock. ²⁶ But everyone who hears these words of mine and does not put them into practice is like a foolish man who built his house on sand. ²⁷ The rain came down, the streams rose, and the winds blew and beat against that house, and it fell with a great crash."" (Matthew 7:21-27)

ABOUT THE AUTHOR

My life is a laboratory. And all human beings are designed as such by the all-knowing God. The only difference among us is that while some willingly become part of life's experiments, some view it from the sideline.

The best lessons we each learn in life comes to us directly and not through a teacher in an academic setting. We all learn and mature in our experiences by trial and error, just like a scientist in the laboratory. But we are not only the scientist, we are also the test specimen and the laboratory facility & instrumentation—all rolled into one.

And when we are in tune with our spirits, it becomes more verification than 'trial and error' because through our spirits, God feeds us great knowledge about our lives, the things around us, and deeper mysteries than we ever thought possible.

Most of my books happened that way. Information came into my mind and takes residence. I soon become aware of it and try to know more about it. As I explore it, it deepens and more is downloaded onto my spirit. And intuitively, I am led to its verification. Once verified, it becomes common knowledge to me.

God has been unbelievably good to me by opening windows to me into great mysteries, such as I have been writing about in my many books. There is hardly a day that I am not writing books. I work on several titles simultaneously, capturing the information as soon as it enters my mind.

Other Titles form this Author:

- Jesus-On – We All Live Because He Lives!
- Who is God!
- What is Love!
- Christ is in Everyone
- Christianity is Life; not a Religion!
- The Singleness of God!
- Overcoming Your Trials!
- Live the Abundant Life!
- Science, Evolution and God!
- Reflections of Life!
- The Rapture, the Tribulations and the Church!
- The Big Bang: and Jesus Christ birthed the Universe!
- **Government & Science: A Marriage made in God's Prophecy.**
- Scientific Proof that the Earth & Water Existed before Time, Space & the Big Bang!
- The God of Science
- Whoever says that Sex is Good is a Liar!
- Revive Your Marriage instantly by obeying God
- Our Universe Rotates inside God's Mighty Hand
- What Is The Truth?
- Every Family on Earth Derives Its Name From God
- LifeBook — Be Strong and Courageous!
- **Human Intelligence is Infinitely-Advanced Cloud Technology**
- Who God Created The Woman To Be
- **Ask What You Can Do For The Kingdom Of God –** Vol. 1
- **Ask What You Can Do For The Kingdom Of God – Vol. 2**
- **Ask What You Can Do For The Kingdom Of God –** Vol. 3
- Life Is A Journey Through God

www.ingramcontent.com/pod-product-compliance
Lightning Source LLC
Chambersburg PA
CBHW020908180526
45163CB00007B/2665